LARGE HADRON COLLIDER

2008 onwards

COVER IMAGE: One of the several experiments at the Large Hadron Collider, the ATLAS detector. *(CERN/Maximilien Brice)*

First published in July 2018

A catalogue record for this book is available from the British Library.

ISBN 978 1 78521 187 4

Library of Congress control no. 2018935485

Published by Haynes Publishing,
Sparkford, Yeovil,
Somerset BA22 7JJ, UK.
Tel: 01963 440635
Int. tel: +44 1963 440635
Website: www.haynes.com

Haynes North America Inc.,
859 Lawrence Drive, Newbury Park,
California 91320, USA.

Printed in Malaysia.

LARGE HADRON COLLIDER

2008 onwards

Owners' Workshop Manual

An insight into the engineering, operation and discoveries made by the world's most powerful particle accelerator

Gemma Lavender

Contents

OPPOSITE A vertical cryostat is tested before being installed as part of the High-Luminosity Large Hadron Collider, which aims to crank up the performance of the collider.
(CERN/Maximilien Brice)

Introduction

‘Scientists Prove Existence of God Particle’. ‘Physicists Find Elusive Particle Seen as Key to Universe’. ‘CERN Scientists Discover Missing Particle’. These were just some of the headlines that led the newspaper front pages on 4 July 2012. Scientists at CERN, the European Organisation for Nuclear Research where the Large Hadron Collider (LHC) is located, had announced the discovery of the Higgs boson.

The Higgs was the most sought-after particle in the whole of scientific history. It had grown into the stuff of legend and had led to the building of a multi-billion euro particle accelerator designed to find it. As far as particle physics goes, the Higgs particle is the glue that binds together the ‘Standard Model’, our picture of the particles that make up the universe. The Higgs, you see, creates a quantum energy field that spans the cosmos, giving fundamental particles, such as quarks and electrons that build up into everything we see around us, their mass.

BELOW The discovery of the Higgs boson made headlines all over the world. *(Alamy)*

20p
NEWSPAPER OF THE YEAR
Wimbledon 2012
Dramatic victory takes Murray through to semi-finals
P52-53
i
Scientists prove existence of ‘God particle’
The essential daily briefing
FROM THE INDEPENDENT
THURSDAY 5 JULY 2012
INSIDE TODAY
Daily codeword
Daily Crosswords
Daily
‘Momentous’ find after 45-year hunt for Higgs boson
Professor weeps as his life’s work finally bears fruit
Physicist deserves the Nobel Prize, says Hawking

As incredible and wondrous as the discovery of the Higgs boson was, the apparatus that discovered it – the LHC – is no less amazing. It’s the most powerful particle accelerator on Earth, and an outstanding feat of engineering. It’s built 175m underground in a gigantic loop on the border between France and Switzerland. Four main giant experiments, and several smaller instruments, are dotted around the 27km circumference, measuring the particles that spill out from titanic collisions between protons, or atomic nuclei, fired at almost the speed of light around a pipe that runs the full length of the excavated tunnel. ATLAS, ALICE, LHCb and CMS – the names of the four detectors that helped make history when the Higgs boson was discovered – have almost become household names. Indeed, the picture of ATLAS that adorns the cover to this book is probably one of the most iconic photographs in science of all time.

But how does the particle collider work? How do you fire countless tiny protons at breakneck speeds with such precision so that they smash together at unprecedented energies, and then detect all the exotic particles that spill out of those collisions? How were the powerful magnets that guide the proton beams installed? What does the signal of the Higgs boson’s existence look like?

ABOVE Peter Higgs stands in front of the CMS detector as it undergoes maintenance. *(CERN/Maximilien Brice)*

BELOW Simulated particle tracks from the CMS detector showing the making of the Higgs boson, which decays into two jets of hadrons and electrons. *(CERN)*

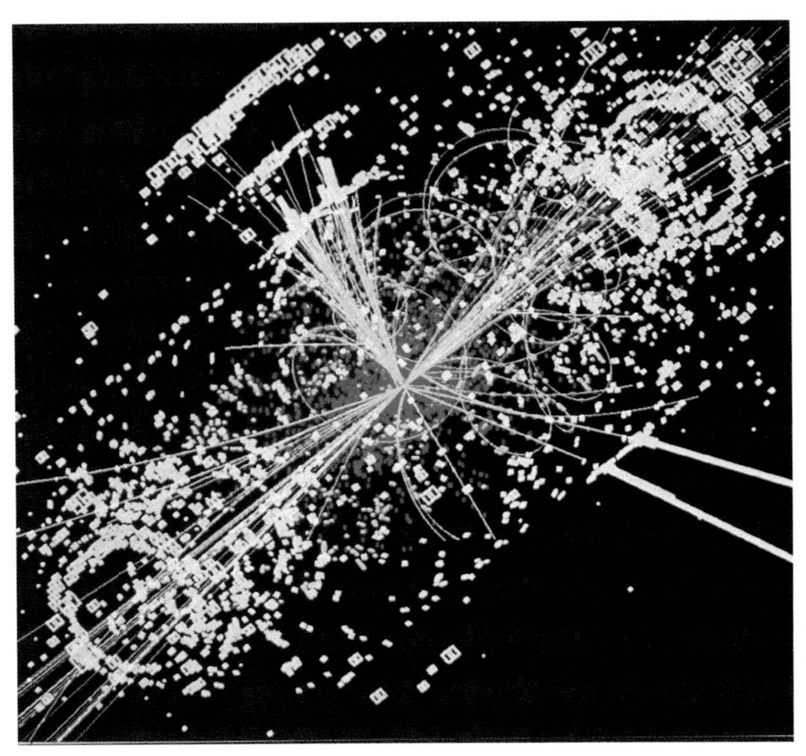

The Large Hadron Collider is one of the most complex scientific experiment of all time. In this manual, we will take you on a tour of the collider, and delve deep into its machinations to understand how they work. We'll explore the physics being probed, and ponder the consequences of its findings, before looking to the future and the next generation of particle accelerators that will attempt to answer the questions that the LHC has kicked up.

The Large Hadron Collider and the discovery of the Higgs boson was simply one of science's greatest adventures. Peter Higgs, who first proposed the existence of the Higgs boson, may have won the Nobel Prize following its discovery, but it was also the LHC experiment itself, and the hundreds of scientists and engineers who designed it, built it and worked on it every day, who are the real heroes of the story.

Gemma Lavender
March 2018

Chapter One

Solving the mysteries of the universe

The universe: a constant source of mystery for us here on Earth. An expanse of galaxies, stars and planets that are forever being surveyed by ground- and space-based telescopes, with every movement in a sea of blackness noted by teams of scientists, working to answer the many questions that it poses.

OPPOSITE Inside the Globe of Science and Innovation at CERN in Geneva, Switzerland, visitors are educated about the Large Hadron Collider's role in understanding the mysteries of the universe. *(CERN)*

But it's not just telescopes that are in on the action: some 175m below ground, in a tunnel beneath the Franco-Swiss border close to Geneva, the world's largest and most powerful particle collider is accelerating particles to breakneck speeds, close to the speed of light, and smashing them together with such force that cosmic windows tightly closed for centuries are being forced apart to reveal new discoveries. The aim of the European Organisation for Nuclear Research's Large Hadron Collider, fired up in 2010, is to look at the universe on a more fundamental level, peering deep into the very basic laws that govern it, unpicking the structure of space and time, and uncovering whether we can really stop the war between quantum mechanics and general relativity by finding the missing piece of the puzzle that unifies them, once and for all.

BELOW Louis de Broglie, the French physicist who first mooted the idea of a giant particle accelerator at CERN in 1949. *(H. Wellcome)*

CERN: home of the Large Hadron Collider

But first, let's meet the physics laboratory that has brought the Large Hadron Collider to life.

We must begin by going back to December 1949, when French physicist Louis de Broglie – famous for his contributions to quantum theory, in particular that matter doesn't just behave like particles, but like waves too, in a concept known as wave-particle duality – put forward the first rough plans for the European Organisation for Nuclear Research. While attending the European Cultural Conference in Lausanne that month, de Broglie reasoned that the proposed organisation wouldn't just unite scientists throughout Europe, but would simultaneously enable them to scratch each other's backs in research terms, which was important at a time when the cost of physics experiments and other such facilities was on the rise. They could pool their funds and share the financial burden, he reasoned.

It was nevertheless almost two years to the day before wheels started turning to establish the European Council for Nuclear Research, and a further two months before 11 countries actually signed up to form the organisation, which was provisionally named the Conseil Européen pour la Recherche Nucléaire, or CERN for short. Admittedly the council got renamed in 1954 to become l'Organisation Européenne pour la Recherche Nucléaire, in English the European Organisation for Nuclear Research, but since the acronym for this would have been OERN, which sounded rather awkward, the name CERN stuck.

A sixth gathering of the CERN council took place during the summer of 1953. Though it only lasted a few days, it made the laboratory an official establishment once and for all. Representatives from 12 countries attended, constituting the founding member states of CERN. These were: Belgium, Denmark, France, the Federal Republic of Germany, Greece, Italy, the Netherlands, Norway, Sweden, Switzerland, the United Kingdom and Yugoslavia (which left in 1961). Since then its membership has increased to 22, the latecomers being Austria, Spain, Portugal, Finland, Poland, Hungary,

the Czech Republic, Slovakia, Bulgaria, Israel and Romania. In addition there are numerous Associate Members and several countries with observer status.

The site chosen for CERN's laboratory and facilities was at the small Swiss village of Meyrin, where construction of its first accelerator, the 600MeV* Synchrocyclotron, was completed in 1957. Since then the laboratory has enabled us to make enormous leaps forward, not just in particle physics but also in benefits to our everyday lives. For instance, the World Wide Web began within the confines of CERN, being the pet project of

* Megaelectronvolts. The energy of beams is measured in electronvolts, which is the amount of energy an electron loses or gains when it moves across an electrical potential of one volt.

ABOVE The flags of some of CERN's member states stand proudly at the physics laboratory's headquarters in Geneva, Switzerland. *(CERN)*

LEFT The 600MeV Synchrocyclotron, which was one of the original instruments at CERN. *(CERN)*

ABOVE The controls to the old synchrocyclotron, which was the first experiment at CERN. They are simplistic compared to the high-tech computers that run the LHC today. *(CERN)*

computer scientist Tim Berners-Lee in 1980, followed in 1990 by Robert Cailliau. Berners-Lee's software was known as ENQUIRE, written in a simple hypertext format with which you'll be very familiar from clicking on all those hyperlinks that take you from one Internet page to another!

Another major CERN achievement was the making of antihydrogen atoms – the 'opposite but equal' versions of hydrogen atoms – which are made up of a positron and an antiproton, rather than the electron and proton used in its naturally occurring state. You'll uncover more of CERN's discoveries in Chapter 4.

As of 2018 CERN is split into five sites: entrances A, B, C and D emerge in Switzerland, and serve as doorways for the organisation's

RIGHT Tim Berners-Lee, who invented the World Wide Web while working at CERN. *(CERN)*

LEFT The Low-Energy Antiproton Ring was a tiny instrument compared to the rest of CERN, and was devoted to decelerating and storing antiprotons for experiments. *(CERN)*

personnel at specific times, whereas entrance E is reserved for those residing in France. There's also an inter-site tunnel, which allows for equipment to be transferred underground – a huge convenience that means no taxes need to be paid as parts of machinery trundle their way along the particle collider's passage. It's also become the home of a whole host of now-defunct and 'ancient' accelerators, such as the Large Electron-Positron collider, the Intersecting Storage Rings and the Low-Energy Antiproton Ring.

The Standard Model of particle physics

Everything we can see in the universe – whether it's a planet, a galaxy, this book or your smartphone – is made up of matter. More specifically, these visible objects are made up of particles: the basic building blocks

LEFT An engineer works on the old Large Electron-Positron collider, which consisted of 5,176 magnets and 128 accelerating cavities in order to accelerate electrons and their antiparticle opposites, positrons. *(CERN)*

RIGHT Everything we can see in the universe is made up of matter – even you. *(NASA/ESA/J.Lotz & the HFF Team (STScI))*

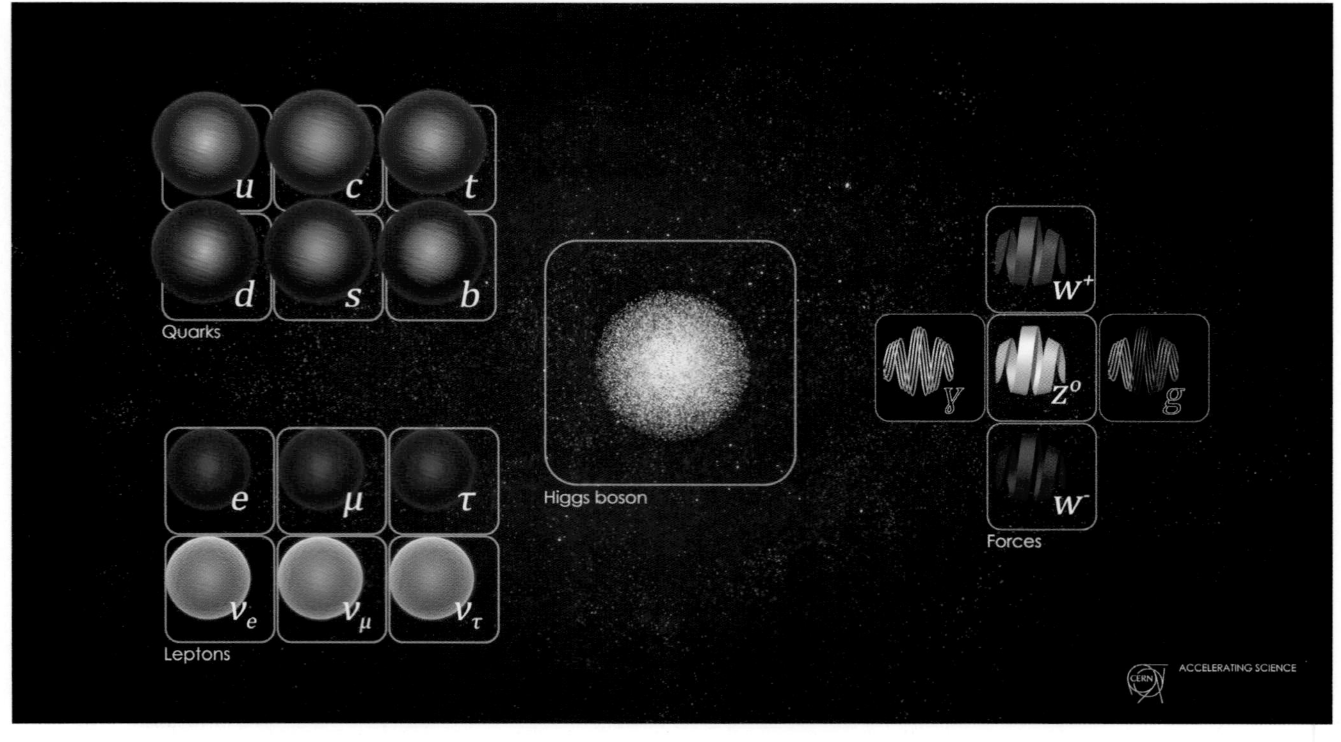

BELOW The Standard Model of particle physics. Quarks come in six flavours (up, down, charm, strange, beauty/bottom and top) and form all of the matter that makes up us, along with the leptons (which include the electron, the muon and the tau particle, and their corresponding neutrinos). *(CERN)*

that supposedly can't be broken down any further. What's more, these particles that make up everything we can observe and touch have fundamental forces that work between them, which decide how these building blocks interact. It's all encapsulated in what's known as the Standard Model of particle physics, which states that particles only occur in two basic types, quarks and leptons; with three fundamental forces at work, strong, weak and electromagnetic.

Another term you'll encounter is fermions. These are particles with a half-integer spin that obey a very important law known as Fermi-Dirac statistics, coined by Italian-American physicist Enrico Fermi and theoretical physicist Paul Dirac. This law governs the possible ways in which particles can be distributed across a set of energy states, with only one particle allowed in each of these discrete states. More specifically, these fermions can't be identical; they have to obey the Pauli exclusion principle, which states that two or more identical fermions can't simultaneously occupy the same quantum state within an atom. All particles have properties called quantum numbers, which describe a fermion's state – that is, the principal quantum number, angular momentum quantum number, magnetic quantum number and spin quantum number – and it's these numbers that need to be unique, to ensure that they're not stepping on the turf of an identical particle.

So, what do these numbers mean exactly? To understand that, you'll have to think of the fermion as an electron inside an atom with a nucleus at the centre of round shells, or orbitals. The principal quantum number describes the energy of each of these orbitals and their relative distance from the electron. The angular momentum number, on the other hand, describes the shape of the orbital; the magnetic number looks at these paths inside a subshell and gives us an idea of their orientation in space; and the spin quantum describes the particle's angular momentum. Spin doesn't refer to the actual physical rotation, yet it gives a magnetic moment as if it does.

Quarks and leptons are fermions. They're also both elementary particles, but there's a fundamental difference between them that sets them apart. The quark is a particle that hates to be alone, so it combines with others to make particles known as hadrons, the most stable of the family being protons and neutrons, which form the nucleus of an atom. Leptons, on the other hand, don't really get involved with strong interactions, and some examples – such as the neutrino – rarely interact at all. Leptons, or more correctly charged leptons, can be found inside the atom accompanying its quark cousins, the negative electrons.

Due to their penchant for interacting, quarks are the only elementary particles to tango with fundamental forces. They also come in six flavours: up, down, strange, charm, top and bottom. Up and down quarks are recognised as the lightest of these; it's this pairing that you'll see heftier members of the family breaking up into, through interactions that bring about particle decay – that's when a particle feels the need to shed some mass. Being at their most stable state means that they're the most common type of quark in the universe, while other, heavier types can be found in the high-energy cosmic rays that shoot across the cosmos, penetrating anything and everything in their path.

The lepton family is also made up of six types, which form three generations: the electronics, muonics and tauonics. When it comes to decaying, the heaviest leptons – such as the muons and taus – aren't that common in the universe. This means that if astronomers are hunting for the most abundant leptons in the cosmos they'll most likely come across electrons. Paired with each electron are electron neutrinos and muon neutrinos, the difference

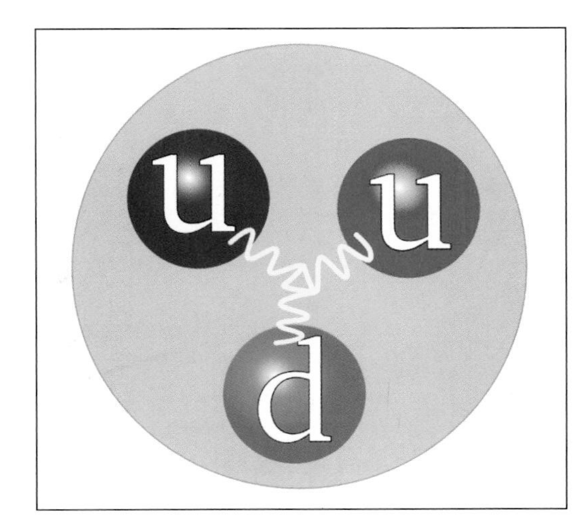

LEFT A proton is made from three quarks – two up quarks, and one down quark. *(Jacek Rybak)*

RIGHT Energy levels of electron 'orbitals' in different atoms. *(Bruce Blaus)*

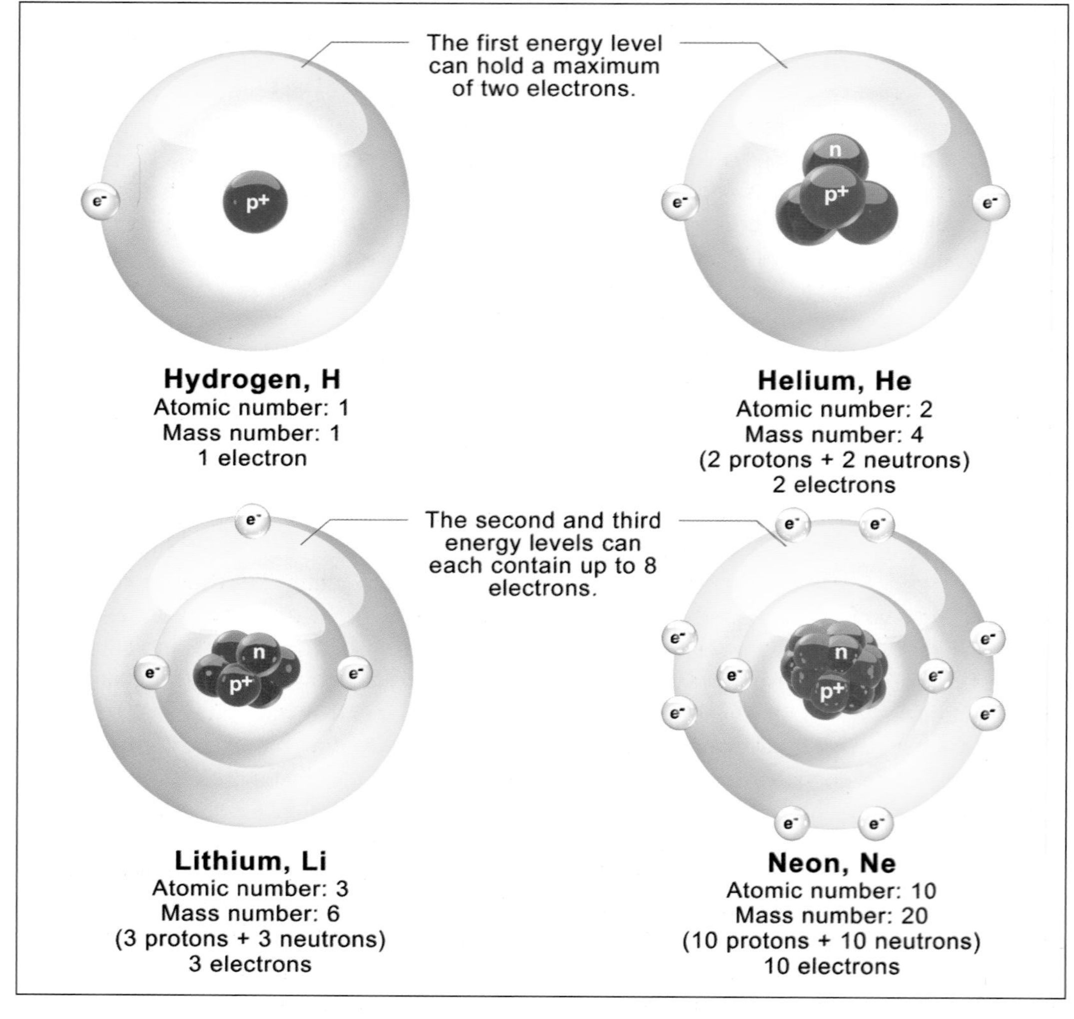

between these being in their charge and mass – the neutrinos are neutral with very little mass whereas the leptons have an electric charge along with a sizeable mass.

Our universe is a place of high-energy phenomena of catastrophic proportions, with four fundamental forces at work between the particles of which it is comprised: the weak, strong, gravitational and electromagnetic. Weak and strong forces work at their best over short distances and can therefore be found between subatomic particles; the electromagnetic forces can be found everywhere in the universe, as can gravity, which, rather surprisingly, is the weakest of the four forces. How these forces come about is the result of a quick exchange of particles known as force-carriers, or bosons. Of these, the gluon is responsible for carrying the strong force, the packets of energy known as photons throw the electromagnetic force for great distances across the cosmos, while W and Z bosons are responsible for the weak force.

But there's a problem. While the Standard Model has served us well since its development in the 1970s, there's a gaping hole in it, with physicists currently ruling that it's not the end of the story: there's a force that doesn't quite make the framework of the Standard Model – an interaction that's familiar to each and every one of us on this planet – and that's gravity.

What's up with gravity?

We're on a hunt for a particle. The graviton is the key to completing the Standard Model of particle physics, and without it the model is incomplete and we're unable to explain how gravity acts on particles of matter. If gravitons do exist, then it's likely that they have no mass and are stable, and a quick wave of your hand will produce them. Of course, given that their energy is so small they're not generating a signal that we can detect.

That's where the Large Hadron Collider comes in. In theory, if gravitons do exist then a particle smasher like the collider would be able to make them inside its chambers, where its detectors would be able to pick up the telltale signs of gravitons at work. Inside the particle smasher particles fly in all directions, creating an atmosphere in which momentum and energy are balanced, and a graviton would leave an empty zone, which we could see as an imbalance of the two, possibly hinting at its existence.

The truth of the matter is that this particle would disappear as quickly as it's made, possibly into extra dimensions, making it difficult for particle physicists to pin down any hard evidence. A graviton would register as a blip in data, which, frustratingly for the physicists hunting them, can be mistaken for all manner of other things. For eight months in 2016, particle physicists debating what could be causing an odd blip discovered that it was being produced by ramping the particle smasher up to its fastest, most energetic collisions. Two different experiments that utilised portions of the Large Hadron Collider – dubbed A Toroidal LHC ApparatuS (ATLAS) and Compact Muon Solenoid (CMS) respectively, which we'll meet again later – seemed to turn up the same results: out of the debris between the crashing of protons, pairs of photons with gargantuan amounts of energy were made. Physicists figured that since not even the Standard Model could explain these extra packets of light, they must therefore hint at new physics or even possibly the graviton – the force-carrier that would finally slot gravity into existing particle physics models.

Alas, it wasn't to be. The extra 750GeV (gigaelectronvolts) generated wouldn't be filling any gaps in our knowledge or resolving any long-standing puzzles. Instead, physicists had uncovered nothing more than pesky neutrinos, getting in the way of a data run. So the search continues, and physicists intent on finding the missing link – whether it takes the form of a graviton or not – are depending on the Large Hadron Collider to help them. Unlocking the secrets of the universe depends on it.

Supersymmetry: linking the bosons to the fermions

We've met the fermion and the boson. These two basic, elementary particles could end gravity's days of being the 'odd force out' by linking it to the other fundamental forces, simply by finding a relationship between their spins. It's easy to fall into the trap of imagining a subatomic particle pirouetting at high speed on its axis when we think of its spin, but this is a misleading analogy – it's a bit more complex than that.

If you were to throw a ball, you'd see it spin in the air before it hit the ground. That ball is free to spin in any direction it chooses. However, though an electron deflected by a magnetic field has spin, or angular momentum, this never really changes, since it's only allowed two orientations – as proven by the Stern-Gerlach experiment, which uncovered that particles own an intrinsic angular momentum by firing a beam of them and observing how they were deflected. It's been established that fermions possess a half spin, while bosons have a full, or integer, spin of 0, 1 or 2 – important characteristics that have allowed particle physicists to come up with a solution to the problem of introducing supersymmetry, a theory which states that for every particle that exists there is a superpartner, or particle that's yet to be discovered, which differs from its partner by half a spin.

According to supersymmetry, then, wherever there's a boson there's also a fermion. And it's not just the spin that makes them different; it's also how they behave around other particles of their own kind. Out of the two, it's the fermions that don't like to be too social – unless they're in a different state or spin – while the bosons are the social butterflies of the particle physics world and prefer to be very alike in their states. They might be different, but supersymmetry brings the two together.

Just to give you a flavour of this idea, supersymmetry suggests that the fermion electron would have a boson superpartner called a selectron, and, in the simplest of supersymmetry theories – where it's 'unbroken' – these superpartners would share a similar mass and quantum numbers. The more complicated version – spontaneously broken symmetry – enables superpartners to have a different mass, and could be an attractive solution to a whole morass of particle physics problems and in particular the Standard Model's hierarchy problem, which, as we've already discovered, is that the weak force is much stronger than the force of gravity.

While supersymmetry could be the answer to the prayers of particle physicists, introducing new physics into the bargain, we mustn't get too excited yet. The Large Hadron Collider is still in the process of finding out if the theory is feasible, seeking out these new particles by creating a high-energy environment. While finding a predicted particle – just like the Higgs boson, for example – doesn't prove that supersymmetry exists, an upgrade to the world's most powerful particle accelerator could provide the lead we've been looking for – especially if we find the Higgs' superpartner, the Higgsino.

The detection of the Higgs boson

During the summer of 2012, it was announced that the Large Hadron Collider had successfully found a new elementary particle that had been suspected to exist as early as the 1960s by six physicists, including Peter Higgs and François Englert. This new recruit to the Standard Model of Particle Physics was the Higgs boson.

The Higgs boson is a very important particle, since it gives others their mass. Being such a big boson, it decays as quickly as it's made, breaking apart into W bosons, Z bosons and photons, which have already been observed in the chambers at CERN.

Without the Higgs, mass – the feature that combines with gravity to gives us our weight – wouldn't exist. (We'll go back to how precisely the Higgs boson was made in Chapter 4.) While its genesis has filled a gap in our understanding of particle physics, the story is far from over, as scientists behind the Large Hadron Collider continue their work to uncover more about the 'God Particle', as the Higgs boson is often called, particularly regarding its properties and rarer decays.

Where did the other part of the universe go?

It's true that dark matter is referred to as the missing part of the universe, but here we're talking about something else entirely. Some 13.8 billion years ago, the universe erupted into existence from a point of space and time called a singularity, which then expanded rapidly, forming the stars, galaxies and everything we see today. But rewind to the first fractions of a second of the cosmos' life and things were very different. The coolness and apparent calmness of today's universe are swapped for an environment that's both exceedingly hot and dense, buzzing with a soup of particles speeding around and crashing into one another.

Look around the cosmos today and you'll know that some of these swift-moving particles must have been matter – after all, everything is made of the stuff. But they weren't alone. Other particles in the mix were antimatter. These are best described as the yin to matter's yang, carrying the same mass as their matter counterparts but possessing an opposite charge – kind of akin to a mirror, where you see an 'opposite but equal' version of yourself in the glass. In terms of particles, for instance, the antimatter version of the negatively charged electron is the positively charged positron.

So, if there's matter then it follows that antimatter should exist too. But there's none around, causing astrophysicists to suspect that something must have happened to tip the balance. If equal amounts of matter and antimatter were made – and going with the theory that for every particle there's an antimatter component – then the universe would contain nothing but energy, given that the two destroy each other when they meet.

To date we've seen particles and their antiparticles undergo sudden transformations that are so quick that they happen millions of times per second before they disintegrate, or decay. It's possible that some entity that we've yet to discover was at work, causing these particles to transform into matter much more often than they decayed as antimatter. Astrophysicists see the early universe as kind

of a balanced system, but that system was disrupted. The mirror analogy implies that the reflection is less than perfect.

Yet, whatever happened all those countless years ago, billions and billions of matter particles managed to survive. And the evidence is in plain sight, from the book you're holding to the surroundings of the room you're in. It's a conundrum that physicists are keen to solve, and it's hoped that they can do so by creating conditions within the Large Hadron Collider's chambers that are similar to a young universe, and then creating matter and antimatter particles by revving up protons to high energies and causing them to smash into each other at super-fast speeds.

ABOVE The Big Bang created the universe as we know it, but the first few moments, where temperatures and pressures were immense, are a mystery that the LHC has been probing. *(LHC)*

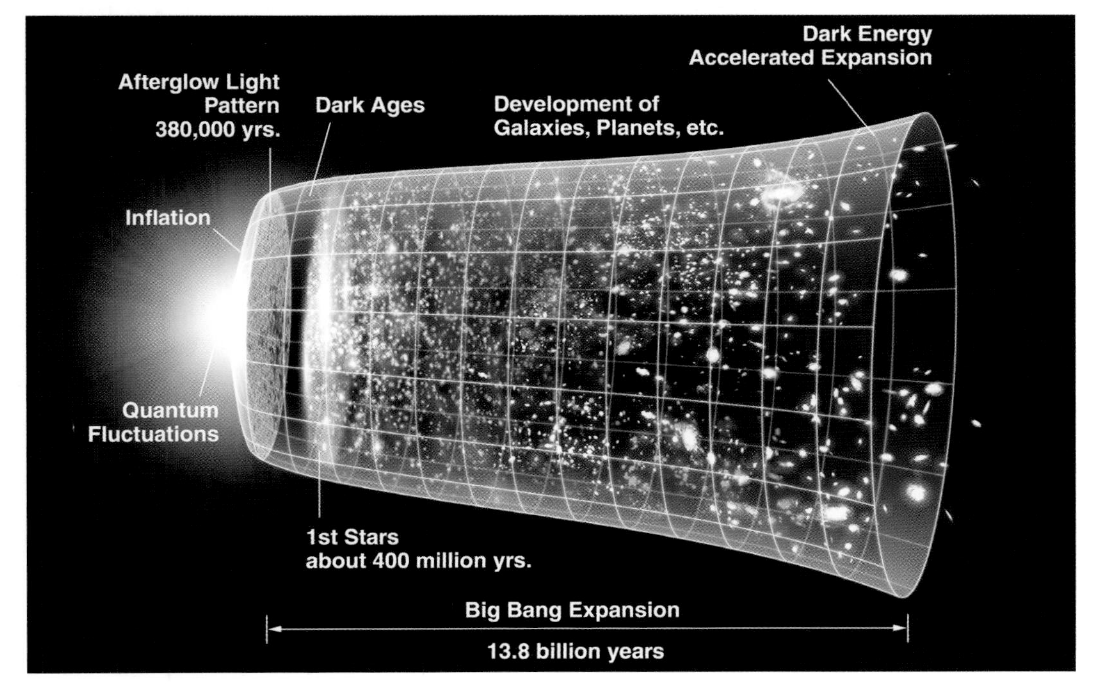

LEFT The history of the universe, from the Big Bang to the present day. The LHC probes conditions similar to those found in the universe just after the Big Bang, as well as the mystery of dark energy that is causing the expansion of the universe to get faster and faster. *(NASA/WMAP Science Team)*

Unlocking the state of a young universe

It isn't just protons that are smashed into each other in the Large Hadron Collider. Some pretty hefty ions also crash into one another under the watchful eyes of its detectors, to enable physicists to determine what matter could have existed in the early universe. It's thought that a plasma permeated the cosmos, a particle soup that's composed of quarks and only tends to exist at very high temperatures and densities. Called a quark-gluon plasma, the hadrons of baryonic matter – or elementary building blocks of matter – aren't strongly attracted to each other due to the extreme conditions.

Roughly ten seconds to 20 minutes after the Big Bang, the universe started to cool down and become less dense. The fundamental forces started to take hold and the bones of elementary particles began to take shape. Fractions of seconds later, quarks and gluons latched on to each other to make simple baryons, like protons and neutrons. It's at this stage in the evolution of the universe that these particles became trapped in this form. Physicists dub this process the confinement of quarks.

Looking back at the basic properties of these particles and how they combine to make ordinary matter is no mean feat. Physicists need to see the quarks when they're free, and that

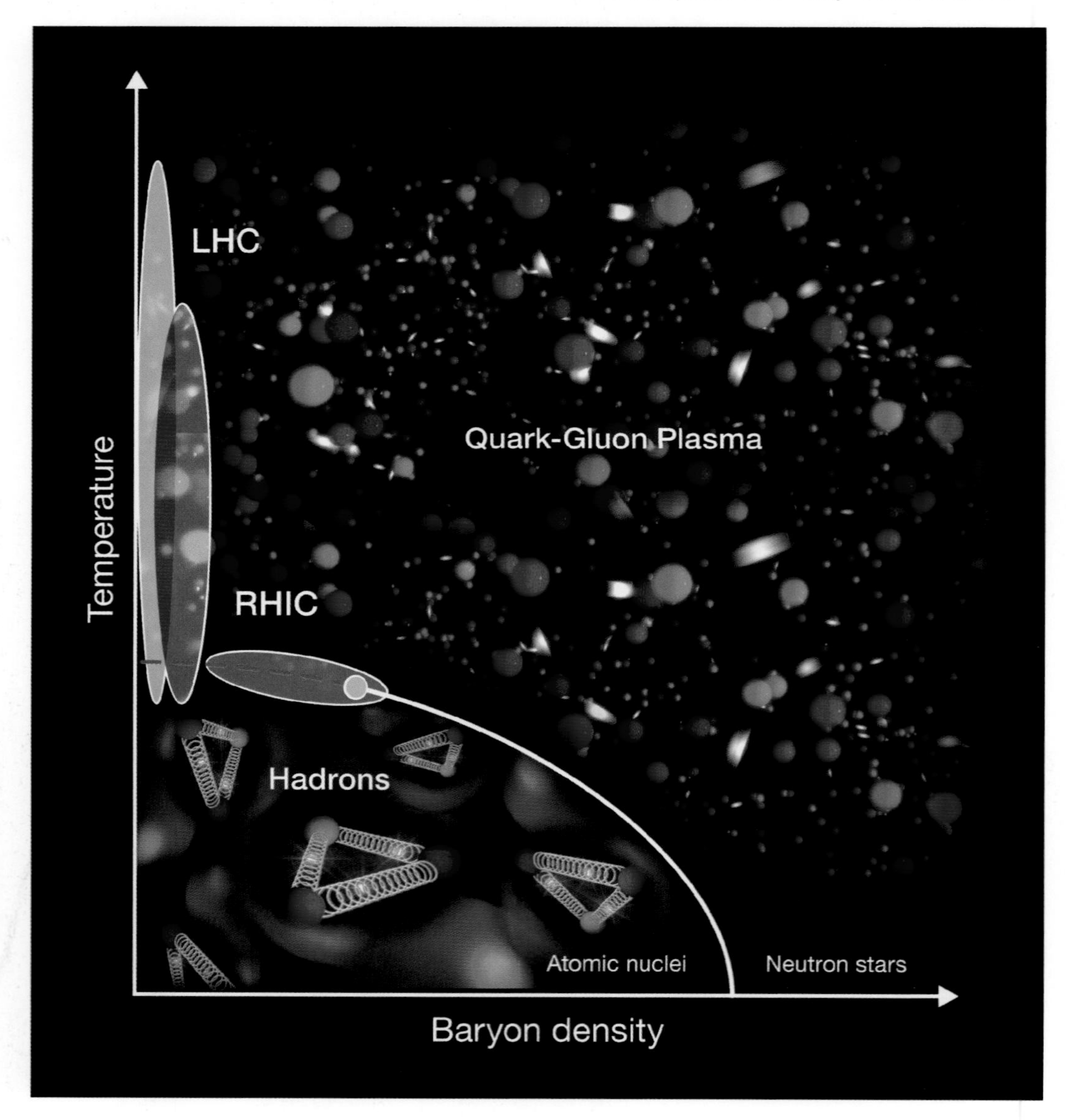

RIGHT A phase diagram showing how matter changes dependent on temperature and the density of baryons – *ie* protons and neutrons. Under high temperatures and pressures a quark-gluon plasma forms. *(Brookhaven National Laboratory)*

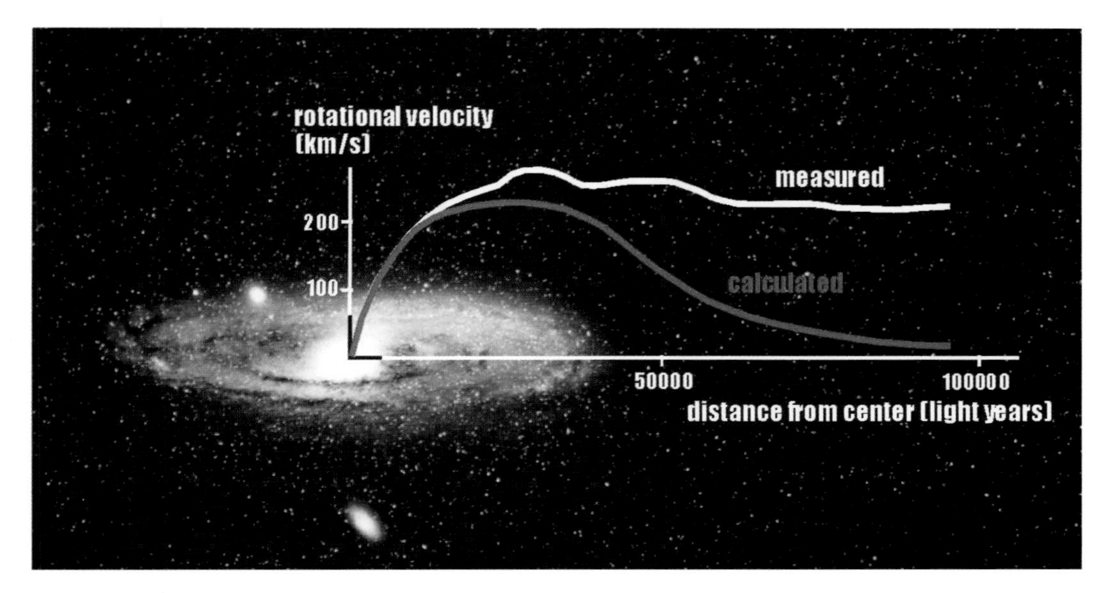

LEFT Some of the strongest evidence for dark matter comes from the rotation curves of galaxies, where stars near their edges rotate just as fast as those nearer the centre of mass. This could only happen if there were some unseen matter on the edges of the galaxies that provide extra gravity. *(NASA/Queens University)*

means momentarily creating areas of white-hot temperature. Heavy ions hit each other, creating an intense 'fireball' that replicates the heat of the new, embryonic universe, exceeding by 100,000 times the temperature you'd find at the heart, or core, of our Sun.

Introducing dark matter

Those first few minutes when the conditions were just right were crucial in the formation of the lightest elements, when the temperature had dropped to a billion degrees Celsius and the density of the cosmos was no more than that of the air we breathe today. Neutrons became attached to protons to make space's deuterium – a hefty form of your standard hydrogen – as well as the nuclei of helium, while many other protons were content to remain single as the nuclei of hydrogen atoms.

So, if this forms the matter we can see, what about the matter that can't be seen at all? As far back as the 1800s such 'dark bodies' were suspected to exist in our galaxy, the Milky Way, because astronomers found the rotation of this collection of stars, gas and dust – as well as other galaxies in the universe – to be extremely odd.

BELOW Based on how the invisible dark matter bends space in and around a giant galaxy cluster (left), astronomers can infer where the dark matter is (right). Can the LHC help shed light on what dark matter may be? *(NASA/ESA/M.J. Jee and H. Ford (Johns Hopkins University))*

RIGHT Approximately 68.3% of the universe is made from dark energy, and 26.8% is made of dark matter. Only 4.9% is ordinary, everyday matter. *(ESA/ Planck Collaboration)*

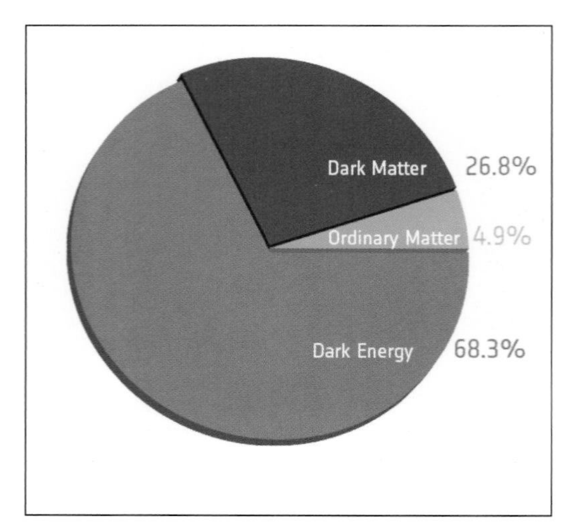

Imagine you're making a pizza. You knead and roll out the dough into a form that's thicker closest to the centre and much thinner at the edge. In other words, you have a pizza base that has density at its centre that becomes thinner as you move towards the edge. After smoothing tomato sauce over it, you put your pepperoni, peppers, ham and cheese toppings on. Finally, before placing it in the oven to cook, you put it on a turntable to trim the edges to make a perfect circle. As you spin your pizza, you might expect the toppings closer to the centre to be moving faster than those at the edge in accordance with Kepler's Second Law of Motion, which expects rotation velocities to decrease with distance from the centre. After all, that's the law that the planets in the Solar System are following the further they are from the Sun. Astronomers expected the same with galaxies; their star and matter-packed centres spin much faster than the stars found at their edge.

But when astronomers looked closely at these structures, they discovered that this wasn't happening at all. Instead they have found that, even distant from the centre, stars' speeds remain consistent. Rather than throw out Kepler's laws, they have therefore reasoned that this odd behaviour can be solved quite easily by adding more mass to the galaxy. So in the case of your pizza you simply add a bit more dough to the base, near the edges, or more toppings – whichever is easier!

The kicker is that we can't actually *see* this matter. It's a mysterious black substance that neither emits nor absorbs light and unsurprisingly, therefore, goes by the name of dark matter. This accounts for roughly 27% of the universe's missing mass. It's best described as cosmic glue that keeps galaxies and galaxy clusters together. Meanwhile, dark energy – a form of energy that's suspected to fill the entirety of space – comprises a more sizeable 68% and works to drive the cosmos. Dark matter tries to keep it together and, possibly, prevent it from tearing apart in the far future.

... but what exactly *is* dark matter?

In addition to the fact that we can't observe it directly, dark matter continues to challenge astronomers in other ways. For instance, they don't quite know what it's made up of. Could it be particles? Or could it be something that we can't even begin to imagine?

The one thing that astronomers do know is that, whatever it may be, dark matter likes to tease us about its existence by pulling a few tricks with visible matter. It may not interact with the electromagnetic force, but it does seem to be quite acquainted with the gravitational force – ironically the only fundamental force out of the four that doesn't fit the Standard Model. One piece of observational evidence for dark matter is something called gravitational lensing. This works in a similar way to how a magnifying glass bends light, allowing us to see fine details up close.

Gravitational lensing uses the powers of Einstein's general theory of relativity by utilising mass to bend light. Objects that hold a lot of mass tend to possess a gravitational field that extends far out into the universe, causing any light rays that stray near it to be bent and focused at another location – and the heftier or more massive an object, the stronger the gravitational field and the more the light gets bent. Observationally, astronomers see this behaviour to a great extent in regions where many galaxies reside, called galaxy clusters, where it's caused by a lump of dark matter slapped between us – observing the universe here on Earth – and the gaggle of galaxies we're looking at. Dark matter might be invisible, but its mass bends the light rays speeding towards us, distorting or lensing them, pulling them into arcs and warping galactic structures to appear much larger than they actually are.

The way dark matter seems to behave could give us some clues as to what it's made up of. It may not be composed of a substance much stranger than dark matter itself and we may not need to disregard the Standard Model in order to understand it. It could be made up of baryons, such as protons or electrons, or it may be that we need to bring supersymmetry into the story, where dark matter is teeming with supersymmetric particles. And astronomers are leaning more towards one than the other.

The MACHOs and the WIMPs

You might be surprised to know that baryons aren't just the building blocks of stars and planets. They also make up more exotic entities such as black holes, objects with such intense gravity that nothing, not even light, can escape them; and neutron stars, the compressed cores resulting from the deaths of large stars. These fundamental particles also play a large part in the make-up of white dwarfs and brown dwarfs. A white dwarf is the ultra-hot husk of a red giant star that has

BELOW The European Space Agency's Planck spacecraft, which has observed the cosmic microwave background radiation and has determined how much of the universe is made up of normal matter, dark matter and dark energy. *(ESA/D.Ducros)*

used up its fuel and blown its outer layers into space, while brown dwarfs are the middle-men between stars and planets but don't really fit into either category, due to their lack of emitted light and their large mass. These objects all have something in common; they don't really emit any radiation and they aren't committed to a planetary system, instead preferring to wander through the universe alone. They belong to a family dubbed Massive Astrophysical Compact Halo Objects, or MACHOs for short.

Galaxies are surrounded by a region that extends beyond their discs of dust, gas and stars. This is known as a halo. A possibility is that these incredibly faint regions, which are suspected to contain dark matter, could in fact be made up of MACHOs, thus making up the invisible mass of the universe. The jury is still out as to whether baryons compose dark matter, but the more observations we make, the more astronomers aren't ready to close that door just yet. Given dark matter's penchant for warping light, the trail of the MACHO is suspected to be easy to follow – especially when a galaxy's halo passes in front of a star, when the star would be brightened by gravitational microlensing; but, to date, hunting for these objects has proved fruitless.

Turning away from their telescopes and surveys of the universe, astronomers have returned to the drawing board, and dissected the idea of baryon-dominated dark matter. Looking back to the birth of the universe, it's unlikely that the Big Bang produced the quantity of baryons that would be needed to produce the volume of dark matter currently understood to account for the cosmos's missing mass. Even turning our attention to the relic radiation that permeates the universe – dubbed the cosmic microwave background – has revealed nothing, as mission scientists behind NASA's Wilkinson Microwave Anisotropy Probe (WMAP) and the European Space Agency's Planck spacecraft discovered that the majority of matter seems to interact with ordinary matter and photons through gravitational effects. This pushes Massive Astrophysical Compact Halo Objects out of the picture.

In a bid to get to grips with dark matter, scientists have gone back to the Standard Model and looked to the Large Hadron Collider for assistance. They figure that supersymmetry could be the key in unlocking the answer to a long-standing question in physics – more specifically non-baryonic matter or WIMPs, Weakly Interacting Massive Particles. The general idea is that for each and every particle in the Standard Model, a partner exists.

WIMPs seem to be the answer. But there's a catch: we've yet to find them – and for that, we need the services of the Swiss-Franco particle smasher. If they exist, they will represent a theory of particle physics like no other. It will be able to make its presence known not only through the medium of gravity, but also through all the fundamental forces that exist throughout space and time.

In theory dark matter particles are quite light, so it follows that we would be able to create them inside the Large Hadron Collider without too much trouble. It's true that they'd escape through the detectors unnoticed, almost as quickly as they were made, but they'd leave a footprint – like the sudden disappearance of energy and momentum straight after a particle smash-up. Particle physicists would be laughing, knowing that they were partway to cracking one of the greatest mysteries of the universe.

The hunt for other dimensions

Look around you. If you were asked how many dimensions you see everything in, what would your answer be? If you were to state that you see your life in three dimensions, then you'd be partly correct. However, if you remember that time is also a dimension you'd come to the conclusion that there are four. Consequently as things currently stand, four is the magic number when it comes to counting the dimensions that compose space and time. The mind-boggling realisation that particle physicists have come to, however, is that we could be very wrong.

As we've already seen, gravity seems to throw some spanners in the works of the Standard Model, because it's the weakest of the fundamental forces. To provide a sense of scale, even a fridge magnet is capable of generating an electromagnetic force much stronger than the gravitational force of planet Earth. This is where the existence of extra dimensions – or superstring theory – comes in. Basically, what if we don't feel the full effect of gravity because it's being spread out in dimensions beyond what we perceive to be space and time?

The father of modern physics, Albert Einstein, pieced together his famous general theory of relativity by 1905, before making final adjustments to his work and completing it for good by 1916. In this theory, he reasoned that space is able to bend, expand, contract and warp in any way imaginable. So scrunch a single dimension to the size of, say, an atom and that dimension disappears. If, however, you could see at such tiny sizes, then that dimension would pop into view again; consequently it could be that there are many such 'strings' in the universe. We just can't see them yet.

The most obvious way to prove that extra dimensions exist is to use the Large Hadron Collider to find evidence of particles that may exist in these dimensions – and they would be heavier versions of the particles you can find in the Standard Model. Particle physicists call these Kaluza-Klein states, and they're where supersymmetry seems to rear its head. If we could get a sniff of the existence of heavy particles that we know carry one of the fundamental forces – for example, a Z or W-like particle – then we could prove that extra dimensions aren't just a figment of science fiction. And if we ever chanced across a graviton, then we would be killing several birds with one stone, while simultaneously opening the door to a whole array of unknown possibilities.

BELOW The Globe of Science and Innovation at CERN in Geneva, Switzerland. *(CERN/Guillaume Jeanneret)*

Chapter Two

Building the Big Bang machine

It was March 1984, and engineers were scratching their heads. They were at a workshop in Lausanne, Switzerland, and they wanted to build a time machine – an experiment that would be like nothing mankind had ever built before.

OPPOSITE Engineers affectionately called this machine 'Taupe', French for 'mole'. The machine was not only powerful, boring through rock at an impressive speed, but was also incredibly accurate, never veering more than a couple of centimetres from its scheduled path. *(CERN)*

It sounded far-fetched to those outside the meeting between CERN and the European Committee for Future Accelerators, but to the attendees there was one obvious way to solve some of the most pressing questions of the universe, and that was to build a 'Big Bang Machine' – an engineering miracle so powerful that it would enable us to look back almost 14 billion years, to the moment when the cosmos rapidly expanded into existence, splashing space and time with the very first fundamental building blocks that form everything we see today. It would effortlessly discover the true nature of dark matter and hunt down the Higgs boson, the particle that's the final jigsaw piece in our model of the universe. It may even uncover the existence of other dimensions beyond the four that are evident to us wherever we look.

But to perform such a feat, the engineers needed a comprehensive plan; something that didn't begin to come to fruition until a decade later. Questions swirled around the thousands of scientists, engineers and technicians who would later be credited with the creation of the largest machine in the world. What particles would they be smashing together? Is it even feasible to get them to collide with the ferocity of a high-energy collision? Where would it even be built?

The big dig

To begin with we must go back to the 1970s, to a point in time when the Large Hadron Collider wasn't even a schematic on a roll of blueprint paper. At this time particle physicists and engineers were considering its predecessor, the Large Electron-Positron (LEP) collider – a circular experiment 27km in circumference, the primary aim of which was to fire beams of electrons and positrons (which we have already seen are categorised as leptons) to energies of 91GeV (gigaelectronvolts) in order to produce the Z boson, a particle which – along with its partner the W boson – is suspected to carry the weak fundamental force of nature.

The late 1970s saw physicists from the member states of CERN considering what the next step should be in their venture to understand the nature of high-energy physics. This was when the Large Electron-Positron

BELOW A technician makes delicate adjustments to just one of thousands of magnets that compose the Large Electron-Positron (LEP) collider, which was fired up for the first time in 1989. *(CERN/Patrice Loïez/ Laurent Guiraud)*

emerged as the experiment that would lead the way, and allow us to build on the results of proton-whizzing accelerators that had barely scratched the surface.

The Large Electron-Positron was formally given the green light by the CERN council during the summer of 1981, but getting the approval of the largest particle physics organisation in the world was only half the battle. French and Swiss teams of engineers had to meet, and the work had to begin. Excavation of where the collider would sit posed the greatest challenge of all. After all, it had to be underground, in a tunnel that linked France and Switzerland and crossed the border between them several times. Constructing a tunnel 100m beneath the surface in the Jura Mountains wouldn't just be an enormous, complicated challenge; it would also be hazardous. If engineers made any errors the whole project would be doomed.

September 1983 saw the presidents of CERN's host countries, François Mitterrand of France and Pierre Aubert of Switzerland, crack open the ground to lay a plaque to

ABOVE Known as LEP-2, this is a superconducting cavity made from the soft, ductile metal known as niobium and was designed as an upgrade to the LEP collider to replace the copper cavities, allowing for beam energies of 100GeV. *(CERN)*

BELOW A radio frequency electric field was used to accelerate electrons and their 'mirror particle', the positron, around the 27km Large Electron-Positron collider. The pictured copper cavities were used to create the 90GeV conditions. *(CERN/Laurent Guiraud)*

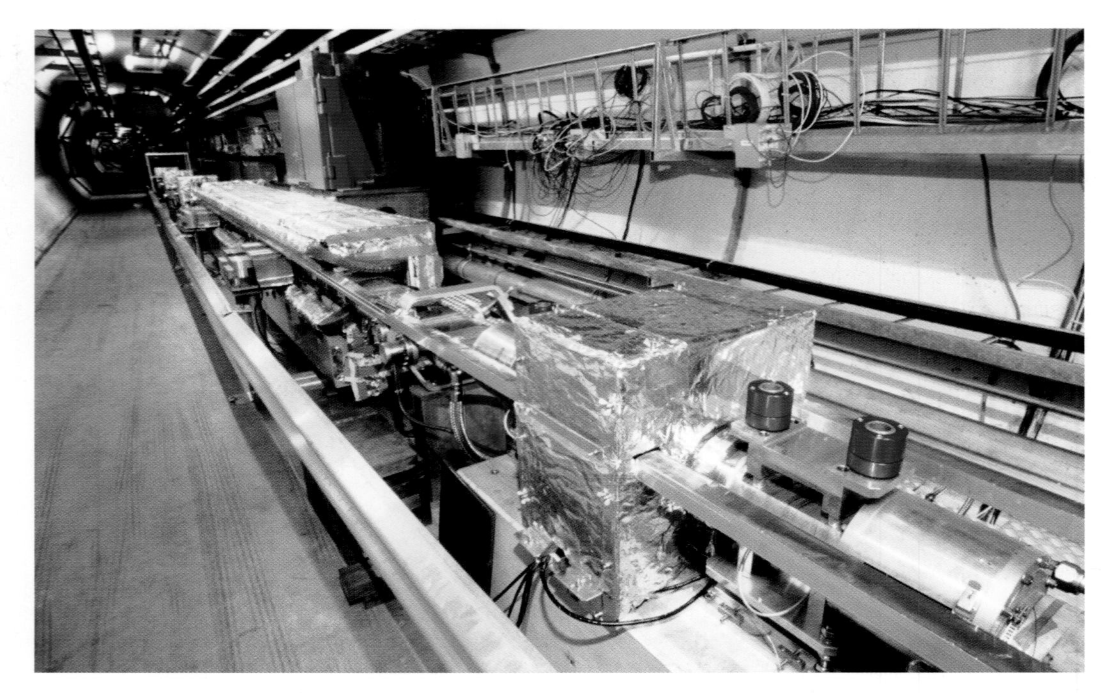

RIGHT A view inside the Large Electron-Positron collider tunnel, which reveals a special dipole magnet – important for measuring the energy of accelerated particles. *(CERN/ Laurent Guiraud)*

commemorate what would be one of the most massive undertakings in the history of civil engineering. It would not be until 1994, when the Channel Tunnel linking Folkestone in the United Kingdom with Coquelles in northern France opened, that the scale of the CERN foundations would be surpassed.

While an impressive piece of man-made engineering in both operation and design, the Large Electron-Positron's cumbersome size meant that red flags were constantly being raised. Geological complications, environmental concerns and political issues all had to be settled before engineering could begin. The tunnel also had to be perfect in shape, to ensure the collider ends met, in order for the LEP to operate effectively. The engineers were only allowed a leeway of mere centimetres – any more or any less and the particle beams would misbehave, firing particles willy-nilly inside the accelerator and ruining any chance of groundbreaking scientific results.

Knowing the geology of the area before the great tunnelling adventure could begin was imperative. Realising that the plain between the 22,686km^2 expanse of the Jura Mountains and the 73km-long Lake Geneva was mainly composed of a solid sandstone called molasse, perfect for tunnels, engineers thought it was an ideal location for their state-of-the-art collider, especially since they'd already plugged the Super Proton Synchrotron (SPS) inside a 6km tunnel of exactly this kind of rock. The engineers therefore reasoned that it would be their safest bet.

But they were wrong. Air pockets in the rock, which weakened the sandstone and rendered it unsafe for excavating, let alone coping with a hefty physics experiment, proved a major bugbear for the engineers as they dug their way beneath the Earth's crust. And there were even greater challenges up ahead, where cracks and faults began to show under the Jura, filled with water due to high pressure caused by thick wedges of rock that reached an elevation of roughly 1,600m above the excavators' heads. The engineering team began to have doubts. That is, until they remembered an earlier exploratory tunnel that stretched 4km from the plain right to the feet of the Jura Mountains. This had been delved to enable the Franco-Swiss team to investigate how sandstone in the plain transitioned to limestone under the Jura.

In order to get more information they now drilled 9km below the surface, before boring three deep holes into the Jura, one of which ended at a depth of 1,000m. What they discovered, particularly at the site of the Jura, stopped the construction of the passage in its tracks. They had come across rocks that were in a bad way, too unstable for tunnelling – Triassic anhydrite salts, weakened rock beds that were 220 million years old, formed when the dinosaurs ruled the planet. On top of these geological faults the team uncovered karstic cavities where limestone had been dissolving, as well as a fault close to the River Allondon where water pressure was 20 atmospheres (the atmospheric pressure you're experiencing right now is 20 times weaker).

All in all they had discovered an underground world that was unfit for the Large Electron-Positron collider. If they were to proceed they faced the risk of even a small accident – where a cavern filled with water at a low pressure – delaying the project by roughly three weeks and costing a further CHF 100,000 (£74,000). The worst-case scenario would be an even bigger accident, where high pressures and water would combine, destroying parts of the tunnel and halting progress by a full year. In addition CERN would face a further CHF 6–7 million (£4.4–£5.2 million) expense to put everything right, and crossing cracks in the rock during the tunnelling adventure would lengthen the project by a further 16 months and incur another CHF 17 million cost (£12.5 million). With time running short and a fixed budget to think about, the team was faced with a hazardous and gloomy prospect – changes had to be made or the Large Electron-Positron project would be over before it had even begun.

Moving the goalposts

The team were not to be deterred. In fact these difficulties inspired them to come up with an improved, more robust plan, and they began to put solutions into action. Underground water reservoirs above the rough molasse were traversed when the access shafts were excavated, while increased water problems for villages in the French part of the Jura, known as the Pay de Gex, would be avoided by having better knowledge of hydrological features in the area and a second-to-none tunnelling strategy. It was well known that these villages were subject to water shortages during particularly dry summers. In order not to exacerbate this situation the engineers therefore made a unanimous decision to move the tunnel from its original proposed location under the Jura Mountains to its present position, several hundred metres further from the peaks and closer to Geneva airport.

Figuratively speaking, their new plan enabled the team behind the Large Electron-Positron to see light at the end of the tunnel. They'd reduced its length from 8km to 3km, they'd avoid three of the most dangerous faults in the Jura, and the quality of rock they'd be digging through was much better. The team also wouldn't need to go quite so far beneath the surface, shaving 450m off the depth of their excavation. They wouldn't be faced with pain- and damage-inducing high water pressure and teams would be able to escape to the surface should any major problems arise. The halls that would be adorned with experiments could also be easily accessed via vertical pits instead of the initially proposed horizontal tunnels, while new ring positions boasted directly available access points – a feature which, as you'll see later, was to be crucial for the operation of the Large Hadron Collider.

On deciding the final resting place of the Large Electron-Positron's tunnel, the situation took a political slant. The land on which the access shafts and their surrounding installation would be constructed wasn't actually owned by Switzerland or France; they needed to acquire it. It just so happens, though, that the French were fortunate in this regard; they already had authorities in place for such an occasion. The Swiss, however, needed to take a vote when it came to the acquisition of land. With a portion of the tunnel in no-man's-land, with the possibility of a Swiss vote going against them, CERN had another trick up their sleeve – they would place the access shafts in regions that would be available to both countries. The organisation had access to the shafts from a previous project, the Super Proton Synchrotron, already situated in Switzerland. All of the others were to be placed in France, only a few metres outside the Swiss border.

The project was starting to come together much more easily, and, rounding up their explosives and cutting machines, the team were ready to tunnel through rock and dirt in earnest. During 1983 and into the opening months of 1984, man and machine worked tirelessly and in unison, chipping away at dirt and sandstone to carve out eight circular arcs plus further straight sections in between which would snugly contain the Large Electron-Positron. Dynamite blew out a total of 18 access shafts to a depth of 150m, which wouldn't just enable teams to transport machine parts and experiments, but were also there to provide escape routes should the tunnel collapse. Underground halls were excavated to house the collider's experiments, as well as additional galleries for the installation of auxiliary equipment. Above the surface a whopping 55,000m^2 of buildings had to be constructed. All in all more than 1.4m^3 million of rock was excavated from beneath the Earth's surface.

The project seemed to be going great guns. But the hard-working excavators were about to hit a snag when they reached what was known as Point 8. Surveys of the sandstone 'roof' here showed it to be sloping towards the airport but covered in 100m of moraines – unconsolidated rock debris originally deposited by glaciers, and so loose that it's a bad idea to tunnel through it. As has been said, the molasse tunnelled through to this point is a solid stone, perfect for tunnels, but even pieces of this sedimentary rock can weaken, particularly when moraines are close by.

Standing on the surface and powering down their digging tools, the team faced a conundrum: they wanted to stay inside the molasse, but they also wanted to be as close as possible to the surface at the foot of the Jura. The solution was to start tunnelling at a slight angle, making the Large Electron-Positron the very first collider to be built on an incline of roughly 1.4%. By excavating the tunnel this way, the team had found themselves a workaround. However, they also left themselves facing several problems, which they realised would come back to bite them later. Principal of these was that though a slight slope in such a large tunnel wasn't immediately obvious, as each engineer walked along it, the two ends of the circle wouldn't quite match up, making installation of the Large Electron-Positron much more difficult. The pressure difference for cooling water was immediately obvious, however, not to mention issues when it came to connecting the Super Positron Synchrotron to the collider that would one day be the most powerful collider in the entire world. But we can deal with these 'minor' problems later, they reasoned.

BELOW **Civil engineering continues to accommodate the Large Hadron Collider (LHC) – in the same tunnel that housed the Large Electron-Positron collider.** *(CERN)*

Meet the machines

How would you construct a tunnel? Many of you will reply that you'd simply get the most powerful digging machine available and, commencing from a chosen starting point, plough through tonnes and tonnes of rock and dirt until you reach the end. However, the team digging the Large Electron-Positron's tunnel went for a different strategy; they would work on it in sections, since, as we've seen already, the ground is composed of different kinds of rock that need to be handled in different ways. The sandstones that the excavators encountered were solid enough for gigantic tunnelling machines with rotating heads to drive through them, boring the tunnel sections with ease and in one go, and, most importantly, without damaging the rock and without shaking the villages on the surface, while still being able to advance incredibly fast underground. Molasse is also very watertight, and won't dissolve when it comes into contact with water. Meanwhile, the section under the Jura was dealt with using explosives, blowing away rock that couldn't be tunnelled through without crumbling dangerously.

Sticking to their allocated working zones,

LEFT Engineers stand with the wagons that assisted the 'Taupe', a powerful machine that created concrete vaults. *(CERN)*

with some of the team working clockwise or anticlockwise towards their 'rubble' excavation point, tunnellers crossed faults and worked uphill inside the passage, keeping the position of the main tunnel synchronised with the intersecting access shafts, surface buildings and auxiliary galleries, advancing a steady 500m per month. If they were to dig the same tunnel by hand they would only have covered 150m during the same time frame.

The tunnelling team were particularly fond of a machine affectionately known as 'Taupe', the French word for 'mole'. Taupe was an impressive piece of equipment, kind of like a small factory in itself. Its head, which weighed in at 170 tonnes, was about 10m in length and about 3.8m wide. Rotating at ten times per minute and packing a punch equal to around 460 tonnes, it was a machine to be reckoned with. A 34-strong set of steel discs worked in unison to cut through rock as easily as scissors cut through paper. While the machine was thrust forwards by means of four cylindrical jacks, two telescopic shields moved alongside the cutting head and temporarily fixed it against the wall, allowing the hydraulic jacks to be anchored. Behind these shields, Taupe cleverly put together the concrete parts of an annular vault even as it continued to bore its way through the rock.

Meanwhile, two conveyer belts were loaded with rubble and other such debris, later to be loaded on to an underground train and taken to the bottom of an access shaft. Trucks waited patiently at the opening, travelling backwards and forwards, emptying and loading, for hours at a time, while beneath the surface workers and machines moved relentlessly forward as teams sealed the excavated tunnel sections with concrete.

Tunnelling through the molasse seemed to be going without a hitch. In fact, progress seemed to be going so well that one machine managed to break a digging record, boring through 59m in a single day, more than twice the average figure. Of course, tunnelling is a risky business, and the workers didn't expect there to be no accidents along the way; rock occasionally worked itself loose from the roof and crashed to the ground, causing teams to scatter to avoid the avalanche. But each time the workers broke through from the main tunnel to one of the shafts they would let out an almighty cheer, slapping each other on the back and taking the time to celebrate before proceeding to the next.

The great tunnel flood

As we've already discovered, it was the ground under the Jura that was going to cause the greatest headache. Workers had to deal with the sections in this region in a delicate manner. Tunnelling under the French peaks would also take the greatest amount of time, as pilot borings had to be constantly drilled into the face of the tunnel to test the geology in order to, hopefully, avoid a major disaster.

Once the workers knew that the rock wasn't about to rain down on their heads, they got to work activating explosives to transform impenetrable walls of limestone and sandstone into curtains of loose rock, 'easy' to clear from the tunnel. This procedure was repeated time and time again as the teams pushed further along their ringed tunnel. But in September 1986 these precautions weren't enough. A geological fault broke apart, falling into the tunnel mere metres ahead of where excavations were taking place. Litres upon litres of water gushed into the passage as the pilot borings struggled to keep it at bay. Escaping from underground, though French workers may have been jocular about the event – calling the leak a Reynard, after the fox in the fable of *The Fox and the Grapes*, with grim jokes about 'chasing the fox' permeating the air – they were as worried as everyone else who watched the water continuing to flood the tunnel. How were they going to continue with the project now?

Engineering teams tried everything they could think of to stem the water flow. They injected resins into surrounding rock, only to find that the continuously gushing water washed the resin away before it could harden. They also injected neighbouring rocks to a depth of 3m, only to be confronted by a second leak, inspired by the first that had got them into this situation in the first place. Nevertheless, the tunnellers were relentless with their injections, never ceasing for one moment from stabbing and jabbing the surrounding walls, while the roof and sides of the tunnel were simultaneously further strengthened by iron and concrete vaults. And at last they'd done it; the water flow began to slow. Their plan had worked.

It took eight months for all the excess water to be drained away – a huge lull in the world of civil engineering. The tunnellers were keen to get started again, and with their new method of injecting the rock and immediately spraying the inner walls with concrete, they could be sure that the landscape wouldn't be shocked by their arrival again. Even more precautions were taken by further strengthening the walls with sturdy, metallic vaults, capable of resisting water pressure. Of course, minor springs of water seeped through the limestone occasionally, but the engineering teams weren't going to be put off this time. That is, until they were confronted by the local fishermen.

Water that leaked from the tunnel might not have looked polluted to the untrained eye, but the sand and fine clay it carried gave it a dirty discolouration that wasn't the most appealing to anyone needing a drink of the clear, fresh liquid generally associated with springs running off mountains. And this muddy water was running into the local river, chock-a-block with an impressive array of fish, thereby worrying anglers, who quickly swapped their fishing

ABOVE Workers cheer at the arrival of the Robbins machine after completing the boring of the Super Proton Synchrotron (SPS) tunnel. *(CERN)*

rods for protest boards, fearing that their game would be affected. Worse still, the French Revolution had inspired a tradition of fishing and hunting amongst the people of France, where it was always seen as a privilege of noblemen. Amused, but by no means willing to upset the locals, the workers agreed that they would clean the water before they released it into the river.

The final push

Looking back at their tunnelling adventure thus far, the 300 people who had risked life and limb to build the foundations for the greatest collider the world had ever known could feel that the end was near. It was time to close the tunnel, where both ends would meet after running the entire length of the circular 27km passageway.

The endpoint was close to Point 4, at the fault of the Allondon, and it would be a delicate operation, where teams would approach simultaneously from both sides. Excavation was continued with gusto, but also due care as pure molasse turned into a mixture of sandstone and clay – a sign that the workers were approaching the underground fracture. Some 80m of tunnel was eagerly stripped away until, when teams on both sides didn't want to push their luck unnecessarily, the group moving towards Point 2 and away from Point 1 stopped boring any further.

A controlled explosion took down the final wall that separated them. Dust swirled through the tunnel, blocking the view that would end the excavation once and for all. After waiting for the dust to settle, a mixture of grinning and concerned faces appeared where a wall had once stood, some breathing a sigh of relief. It was February 1988 and the tunnel was complete.

The LEP's accomplice: the Super Proton Synchrotron

When it was built, the Large Electron-Positron (LEP) served to continue the work of a European sequence of colliders such as the Italian-based Anello Di Accumulazione (AdA) and ADONE (Big AdA), France's Anneau de Collisions d'Orsay (ACO) and Germany's Doppel-Ring-Speicher (DORIS). As is apparent from its name, the Large Electron-Positron was an electron-positron collider too, and was inspired by and would continue the work of its fellow accelerator machines long after their accelerator beams had been switched off for the last time.

Just as there are gaps today in the Standard Model of particle physics, there were more leading into the 1970s. By 1976 physicists were still unsure which particles were responsible for carrying the weak force, while the subatomic particle, the top quark, was yet to be discovered. The team that would lead the LEP believed their machine would be revolutionary – after all, the Positron-Electron Tandem Ring Accelerator (PETRA) in Hamburg, Germany, had started up in July 1978, firing particle beams to energies of 19GeV, while another collider based at Stanford Linear Accelerator Center (SLAC) in California, and known as the Positron-Electron Project, or PEP for short, was due to be started up any day. Both would surely make groundbreaking discoveries that the Large Electron-Positron could follow up; or better still, LEP could fill gaps in our understanding of the Standard Model.

BELOW The Large Electron-Positron's project leader Emilio Picasso addresses representatives from CERN's Member States at the inauguration ceremony of the newly installed collider. *(CERN)*

LEFT The memorial plate that commemorated breaking the ground for the Large Electron-Positron collider on 13 September 1983. *(CERN)*

Before the LEP could even fire its first particle beam, scientists at CERN had made an announcement; their antiproton-proton collider, originally dubbed the Super Proton-Antiproton Synchrotron and later renamed the Super Proton Synchrotron, had made a huge discovery. A mere couple of years after it was first switched on it had found the two flavours of boson – the same W and Z bosons that are responsible for carrying the weak force.

It's important to realise that the Large Electron-Positron was an ideal machine, perfect in producing W and Z bosons and measuring their properties. It was the CERN community that became impatient – they couldn't wait for the construction of their prime project; they wanted another accelerator to be made, one that wouldn't jeopardise the Large Electron-Positron and at a mere fraction of the cost. And so the Super Proton Synchrotron became their pet project, made by modifying a pre-existing proton accelerator already running at the Fermi National Accelerator Laboratory (Fermilab), close to Chicago in the United States, into a proton-antiproton smasher.

As you'll hopefully remember from Chapter 1, antiprotons are known as the 'mirror particles' of positively charged subatomic protons, their major difference being that they're negatively charged. With this in mind, the Super Proton Synchrotron would be made with simply a single vacuum chamber, very unlike a collider that smashes protons head-on at high speed in an environment where magnetic fields speed in opposite directions and where several chambers are needed. Instead the single 'drum' of the synchrotron was enough, the particles' opposite but equal charges allowing them to navigate the same magnetic field in opposite directions and into head-to-head smash-ups, where they annihilate themselves. These are conditions in which W and Z bosons are made with ease.

You might be wondering how these negatively charged particles can be made, let alone stick around long enough to even exist. To achieve this, CERN used a bit of help from an infrastructure that had been in existence at the centre of scientific research since 1959, the Proton Synchrotron (PS), coupled with the eloquently named Antiproton Accumulator (AA). Antiprotons were made by directing an intense proton beam on to a target from the PS. Bursts of the subatomic particles came thick and fast, before being steered towards the accumulator, to be stored there for several hours.

Metaphorically stacked to the rafters inside the Antiproton Accumulator, the Proton Synchrotron and the Super Proton-

ABOVE Engineers install dummy dipoles in a mock-up of the Large Hadron Collider's tunnel. *(CERN/Christoph Balle)*

Antiproton Synchrotron would get ready to be filled, where three sets of 100-billion proton bunches were accelerated to 26GeV in the synchrotron, and later injected into the proton-antiproton collider. Next, roughly 10 billion antiprotons also shot into the synchrotron from the opposing direction. These too were to be injected into the Super Proton-Antiproton Synchrotron to be accelerated to 315GeV. Lying in wait, the Underground Area 1 (UA1) and Underground Area 2 (UA2) detectors didn't miss a trick; all of the annihilation action – which met at six points – was caught at their very centres, registering any results. The process would then start over.

With the pressure to uncover the bosons forever hot on the heels of the CERN scientists, the Antiproton Accumulator wasn't accumulating antiprotons fast enough, sometimes taking several days to create a batch of antimatter particles. CERN needed a device that would increase the accumulation rate ten times over, and their prayers were answered with the construction of the Antiproton Collector, which was built around the accumulator between 1986 and 1988. It improved antiproton production by a factor of ten.

Despite being knocked together in almost no time at all, the Super Proton Synchrotron became the workhorse of CERN's particle physics programme the moment it was switched on in 1976. It was originally designed to accelerate particles to 300GeV, but its sheer structural engineering helped it to smash that record, bumping it up to the capabilities of a 400GeV accelerator. Straddling the border between France and Switzerland and 6.9km in circumference, it became an integral part of the Large Electron-Positron, serving as an injector to accelerate electrons and positrons to almost the speed of light before spitting them into the synchrotron's ring. Without it and its 1,317 electromagnets and 744 dipoles to bend particle rings into circular paths we might never have been able to probe the intricate details of protons or probe into the absence of antimatter in the current state of the cosmos.

What the LEP is made of

Before we head back down into our newly excavated tunnel we're going to take a moment to look at what a particle collider such as the Large Electron-Positron constitutes, and how its components helped it to break down some of the seemingly impenetrable walls of particle physics. You'll read all about such discoveries later on in this chapter.

Proton-proton colliders were a messy business, with scientists needing to spend a great deal of time making head or tail of the results they produced. The Large Electron-Positron was a clean machine in comparison, the electron-positron patterns it made inside its vacuum chamber being easy to analyse and decipher, thereby introducing a whole new level of accuracy that would pay dividends in future particle physics. This enabled massive, groundbreaking strides to be taken.

But in order to make these leaps and bounds, the Large Electron-Positron needed particles to play with. That's where CERN's accelerator complex comes in, a succession of machines that send subatomic particles flying to higher and higher energies. Even before they make their way into the Large Electron-Positron's very large accelerator ring to be whizzed around at breakneck speed, hitting the detectors as they go, the journey of the electron and positron is actually quite a complex one.

Firstly, you obviously can't have a particle collider without particles, and in the case of the Large Electron-Positron electrons need to be made. Particle physicists use the magic of thermionic emission to make them – that's where these negatively charged particles are boiled off a thin wire filament that's heated to high temperatures by an electric current. The device, which you can actually find in your television, is called an electron gun. Its task is to then knock these electrons off into the vacuum of space, and an electron beam speeds from whence it was generated. In your TV this gun creates the picture of your favourite television programme. At the Large Electron-Positron it meant that the collider was just getting started in its data run.

Within a split second of being made, the electrons make their way to the Large Electron-Positron Injector LINAC, or LIL for short. The

BELOW These sextupole magnets, which are arranged by alternating north and south, work in unison applying corrections to the spread of the high-energy beam and also for the Large Electron-Positron, to achieve better focusing. *(CERN)*

ABOVE Engineers get stuck into excavating one of the transfer tunnels that would later be used to provide pre-accelerated protons from the Super Proton Synchrotron (SPS) to the Large Hadron Collider (LHC). *(CERN/Laurent Guiraud)*

main aim of LIL's game was to also create positrons, which it did quite effortlessly by smashing electrons into a tungsten metal target. The positive particles, and what remained of the electrons, were then passed on to the Electron Positron Accumulator (EPA) ring. LIL was always reliable, producing an abundance of particles at energies of 500MeV. But these particles didn't yet have the high energies in which scientists were interested.

The EPA helped, and it wouldn't let go of its collection of electrons and positrons until they were ready. The accumulator served as a buffer between the fast-cycling 'upstream' LIL and the much slower 'downstream' Proton Synchrotron and Super Proton Synchrotron, the latter of which we met earlier on in this chapter. In turn, they would accelerate the particles to 28GeV and 450GeV respectively before injecting them into the particle-antiparticle collider's major ring, which had a total of eight highly circular arcs and four straight sections.

It's quite easy to envisage the inside of the Large Electron-Positron's circular 'accelerator' as an empty tube, only filled with particles after being injected by the Super Proton Synchrotron, when they make their way round and hit each detector as they pass. However, despite the speeds at which they're thrown through the Large Electron-Positron's major ring, these electrons and positrons still need a helping hand to make them go even faster, notching up speeds close to that of light. To do so, 31 magnetic cells were ready and waiting to guide the beams around each of the arcs that were packed with incredible technology – focusing and defocusing quadrupoles for the magnetic field, vertical and horizontal orbit correctors, focusing and defocusing sextupoles and a group of 12 bending dipoles, capable of bending the beams around. Each cell was roughly 79m in length to help the electrons and positrons on their way. It was in the straighter sections of the ring that the particles went hell for leather. This was also where the detectors laid in wait, electrons and their antiparticles using them as checkpoints on a racetrack, and where the results of collisions would unfold.

The Large Electron-Positron's detectors were almost like great, complicated institutions in themselves. They've also been likened to onions, with each of their 'skins' or layers serving a very important purpose. These experiments were able to cope with the high energies manifested by the collider and the high level of precision required of them. Their sizes were also impressive, being many metres high and long and of massive weight, tipping the scales at thousands of tonnes.

The four that composed the Large Electron-Positron might have been different, working in a complementary fashion, but they also shared some similarities – at least in terms of their design. The particles produced in the head-on collisions would crash into each other with such speed that they flew off in a variety of directions, so if the detectors weren't large enough then they would surely be missed. Neutrinos, hugely abundant particles which are similar to the electron but without the charge, are slippery customers when it comes to using detectors to uncover their arrival in particle accelerators. They tend to appear when the weak interaction is lurking – the shortest of the fundamental forces. If they have mass, then these particles would also interact with massive particles, something we can't know for sure until we learn more about the weakest of the four known forces – gravity.

Detectors the size of those on the Large Electron-Positron added fuel to the fire in the hunt for particles predicted by the Standard Model; we just hadn't seen them yet – at least, not directly. Some particles were considered to be traceless, so the teams behind the detectors found a workaround: they would hunt for the missing quantities they left behind. It sounds complicated, but by seeking out properties such as energy and momentum, or, rather, their gaps in an otherwise uniform field, they could infer their existence. Of course, engineers had to ensure that their experiments were enclosed, with no cracks in their structure through which particles could escape. Such carelessness would ensure that observations were compromised and their exploitation of the laws of conservation scuppered.

The Large Electron-Positron's experiments went by the names of the Apparatus for LEP Physics (ALEPH); the DEtector with Lepton,

BELOW Diggers pull away rock and dirt from one of many detector sites of the Large Hadron Collider. *(CERN/Maximilien Brice)*

40 TON

OPPOSITE The Detector withr Lepton, Photon and Hadron Identification, or DELPHI for short, was one of four experiments of the Large Electron-Positron collider. It has since left its cavernous home at Point 8 and is now on display at CERN. *(CERN/Maximilien Brice)*

Photon and Hadron Identification (DELPHI); L3, because it was proposed in the third Letter of Intent regarding the LEP, submitted in 1983; and last, but by no means least, the Omni-Purpose Apparatus for LEP (OPAL).

While all experiments were gigantic, L3 was the largest. Its octagonal magnet was 14m long and an impressive 15.8m high, making it roughly nine times taller than the average human. Weighing in at 8,500 tonnes, this mega-magnet was responsible for the experiment's mind-blowing precision. L3's major role was to identify and figure out the speed of movement, or momenta, of packets of light known as photons as well as leptons. For this it employed a brilliant electromagnetic calorimeter made up of 12,000 bismuth-germanium oxide crystals, produced from a very purely grown source of hard, grey-white metalloid crystals at the Shanghai Institute of Ceramics. Photons and electrons become trapped in these crystals and, being excited, let off electromagnetic cascades that are picked out by the calorimeter, which witnesses light flashes. The brighter the flurries of light, the more energy the particle has emitted. This piece of apparatus was then encased in a hadron calorimeter, weighing in at 400 tonnes and made of depleted uranium, a silvery-white metal, that was fashioned into plates for second-to-none precision in picking out cascades of hadrons.

As you will discover, L3 wasn't like the other experiments. Its magnetic coil sat close to the edge of the detector, allowing it to track speeding muons – particles that use their charge to interact with matter and, being leptons, also only interact via the weak interaction, using very large wire chambers measuring 5.5m long and 2.2m wide. This was the 'barrel' system, which allowed for a high precision for about 30m – an impressive feat given the chambers' large size. It's known that muons like to tunnel deeply into materials, knocking electrons off the atoms that

ABOVE An end-view of the Large Electron-Positron's 10-metre-diameter DELPHI detector. *(CERN)*

LEFT A single magnet, which encased the Large Electron-Positron's L3 detector. These magnets formed the experiment's characteristic octagonal magnet, currently being reused by the ALICE detector. *(CERN/Serge Dailler)*

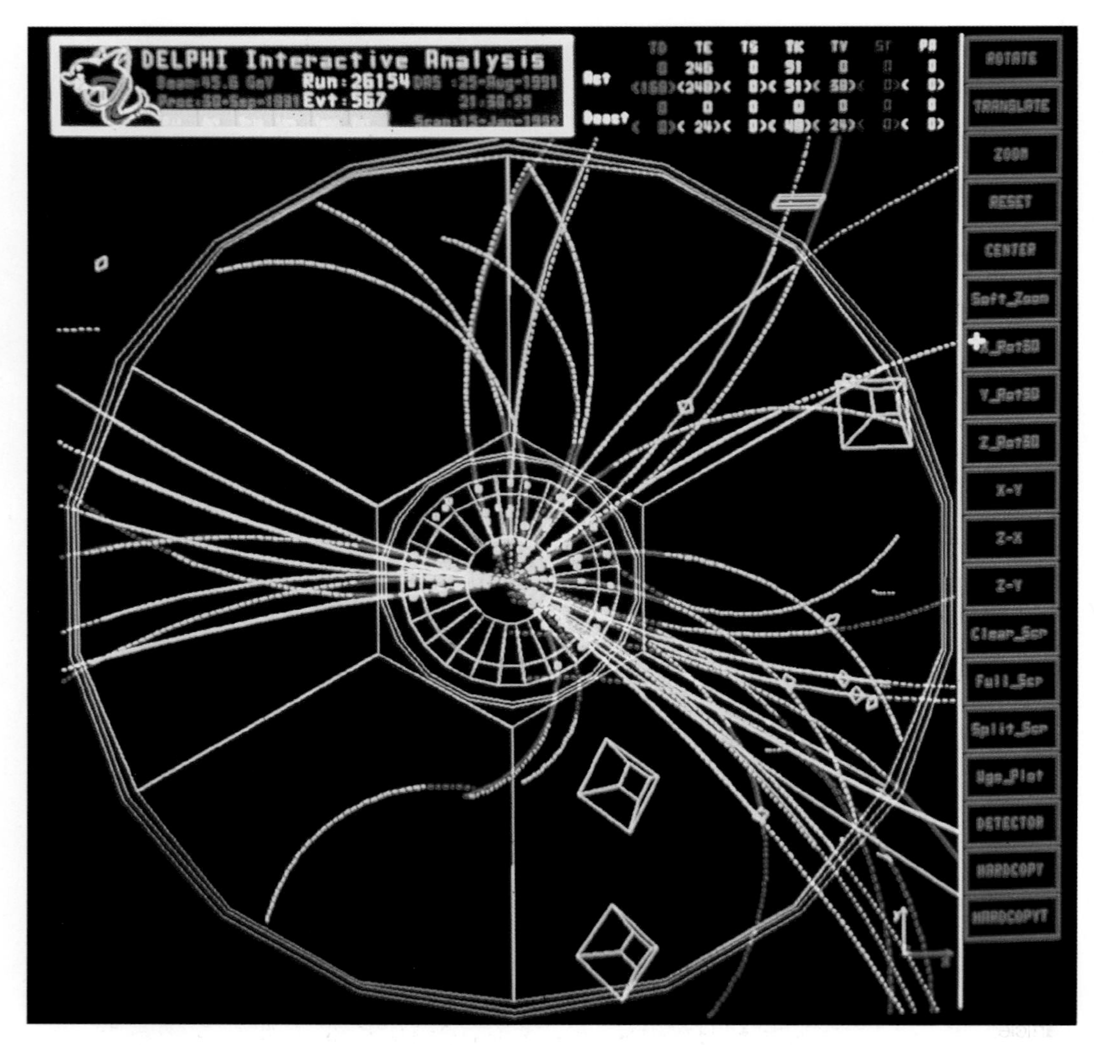

RIGHT Real data obtained from the Large Electron Positron collider's DELPHI experiment. Three jets can be seen, the telltale sign of a Z boson decaying into a quark and antiquark, along with the emission of a gluon. The LEP sees the jets as showers of hadrons. *(CERN)*

make up the matter, hence ionising them and giving them a charge in return. Compare muons to electrons, though, and you'll discover that they don't produce very much electromagnetic radiation to detect, hence the need for the high sensitivity of the L3 experiment.

While L3 hunted for hadron flashes, DELPHI was given the task of identifying hadrons and getting stuck into figuring out the characteristics of fellow subatomic particles, the leptons and the photons. In order to do the task justice, DELPHI was fitted with two counters, while at the heart of the experiment Time Projection Chambers (TPCs) traced the particles' tracks as they whizzed through the LEP. These detectors were found inside a 6.2m-wide and 7.4m-long superconducting solenoid, an electromagnet powered by the coils of superconducting wire. At the time, DELPHI's solenoid was the greatest magnet of its kind, between 18,461 and 48,000 times stronger than the magnetic field that encompasses our planet. Meanwhile, the hadron calorimeter and muon tracker – features of L3 – sat outside the coil, lying in wait for its particle-type quarry to pass by.

Moving anticlockwise along the LEP's circular tunnel you'll meet OPAL. This experiment was expected to be ready as soon as the particle collider was powered up, for it used much more conventional techniques in its operation, techniques that had already been employed on the JADE detector on the PETRA collider in Hamburg. OPAL sought out the locations of charged particles with the help of a jet chamber, pumped heavily with gas, while some 4,000 wires reached vertically from wall-to-wall. How OPAL worked was dependent on a particle smashing into a wire; when this happened, an electric signal was transmitted through the metal, carrying the coordinates of the particle's location.

To ensure that the LEP was complete, the ALEPH experiment was also included. Unlike OPAL, which was considered a 'safe experiment', this general-purpose detector was hailed as an ambitious machine, capable of managing anything physicists threw at it. Its central chamber, just as with DELPHI, was a TPC. At the time TPCs were considered a risky piece of technology, and confidence that ALEPH would be able to measure particle tracks due to its 4.4m length and 3.6m cylindrical width was questionable. But on packing the component with a mixture of argon and methane gas, engineers thought it was worth a try, and they soon found that it was a risk worth taking – particles inside the detector could be seen being dragged along by an electrical field some 20kV per metre in strength. Along the way these fundamental particles would smash into gas atoms, freeing their electrons, which would then drift towards and crash into end plates some 10m^2 in area.

This set-up enabled physicists to reconstruct the particle tracks in three dimensions, a feat that wouldn't have been possible without the help of ALEPH's superconducting magnet, marginally stronger than the one hosted by DELPHI. This component gave charged particles their momenta, as well as bending their path in the TPC.

These detectors wouldn't just be an engineering marvel: they would be the key to determining the properties of particles and would be essential in determining the design of the Large Hadron Collider, as will be seen.

Piecing together a Large Electron-Positron collider

Just one thing stood in the way before completion of the Large Electron-Positron's underground chamber could be celebrated – a blue ribbon, which was promptly snipped amid a round of applause. The 300 or so engineers responsible for the tunnel's construction took part in a ceremony inside the underground hall of Point 4, some 150m beneath the surface, which formally symbolised the end of what was, at the time, the biggest dig in history. Then it was time to place the Large Electron-Positron inside, and the engineers had to be precise. No deviations could be accommodated en route, with the collider's storage ring – eight circular arcs joined by a long straight section – needing to fit snugly within the newly excavated 27km tunnel.

But the underground passage covered an impressive amount of land, so the team needed help. Guiding the components of the collider into position had to be turned into an art of impressively accurate proportions, with wiggle-room of less than a millimetre if it could be done. Enter Terrameter, an instrument developed in the United States for earthquake research, and Navstar, operated by the United States Air Force, a global positioning system in Earth-orbit which during the 1980s was in its infancy but already mapping out the gravitational field of our entire planet. It was the very first time that either system had been used on European soil.

Together, Terrameter and Navstar worked effortlessly, the former employing two laser beams of different wavelengths – such as blue (0.4416m) and red (0.6328m) – to ensure that wobbles in temperature and pressure wouldn't scupper achieving an accurate alignment, while the space-bound constellation of satellites managed to seek out a handful of points which would need to be surveyed every few years. Due to the enormity of the Jura, engineers feared that its mass would noticeably distort the terrestrial gravity, with the land becoming unstable. Indeed, a slight shift of 2mm over several years was noted; not a problem if the LEP was built above the surface, but an added layer of complication when built underground. To get a clearer picture of the collider's positioning and to help in navigating its components through the extensive passages, these reference points had to be located deeper underground whilst simultaneously taking everything else – including the planet's curvature – into account.

With an impressive length of 27km, well over 5,000 magnets and 128 accelerating cavities, the Large Electron-Positron was huge and quite a challenge to transport through the tunnel, an aspect that would also rear its head when

ABOVE Inside a clean room, engineers are sufficiently suited and booted in their overalls as they get to work on one of the LEP collider's superconducting cavities, allowing high electric fields to be produced and the generation of higher beam energies. *(CERN)*

RIGHT The Large Hadron Collider (LHC) occupies a circular tunnel below the Franco-Swiss border. Look down the passage in this image of the LEP's housing, and its huge circumference makes it appear almost straight! *(CERN)*

it came to operating the particle collider in the future. Transportation of men and machine parts alike needed not only to be efficient, it also needed to stick to a schedule, since very few access shafts were available and the engineers generally needed to walk several kilometres to reach the places where the collider was being pieced together.

Needless to say, the engineering teams weren't overly fond of their monotonous marches down the tunnel, helmets crowned with a central light bulb that threw out as much light as a single candle in a vast cave. On reaching their work stations they would stop for a rest, crouching down in the concrete layered tube that they'd excavated just months before and having a breather before resuming another day's work. They figured out that a huge proportion of their working day was spent just travelling to and from their workplace, descending one of the available shafts and then trudging to wherever they were working, which meant covering a mere three miles per hour at average human walking pace.

The unanimous grumbles of the workers were eventually noticed by CERN, who decided that the solution was to install a monorail system with its trains suspended from the tunnel roof, a design not too dissimilar to the trains that pull into and head out of stations all over the world every day. Two cabins for drivers capped several wagons – which differed for those carrying workers and those transporting parts for the Large Electron-Positron – at both ends, their motors capable of pulling an impressive 7.5 tonnes. With different kinds of monorail train assigned to carry specific cargoes, scheduling was key. To CERN, the monorail represented money well spent. Not only did it minimise complaints, it also ensured that 60,000 tonnes of collider components were locked in place within just 18 months.

The collider's very first magnet was installed ceremoniously under the watchful eyes of

BELOW Four Large Electron-Positron (LEP) monorail trains were suspended from the ceiling and transported materials and workers the length of the 27km-long tunnel. *(CERN)*

RIGHT The LEP's first magnet is installed on 4 June 1987, as part of a ceremony attended by French Prime Minister Jacques Chirac and Swiss President Pierre Aubert. *(CERN)*

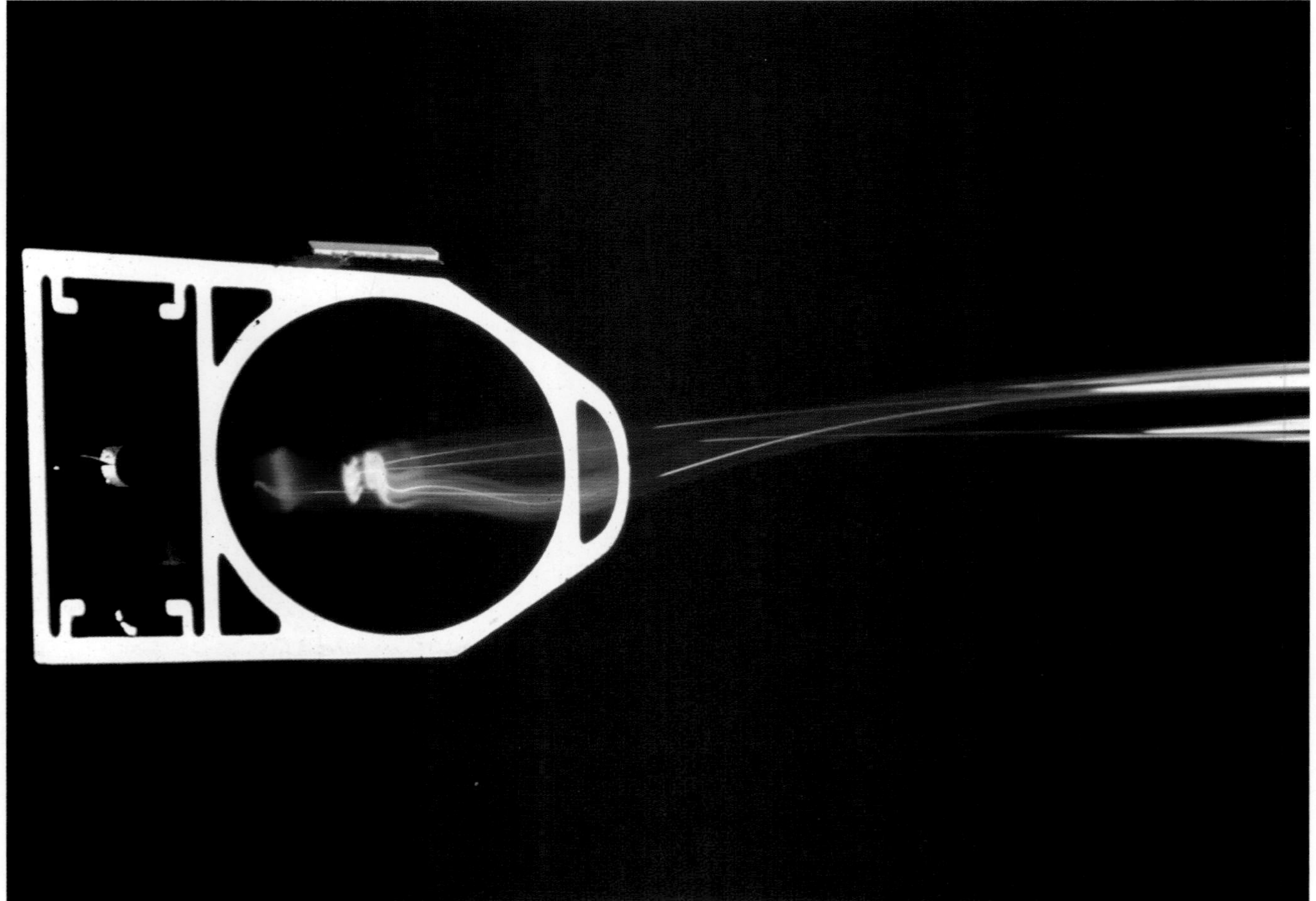

BELOW A CERN illustration, which shows how beams move through the vacuum chamber of the Large Electron-Positron collider if we were able to detect them with our eyes. *(CERN)*

Prime Minister Jacques Chirac of France and President Pierre Aubert of Switzerland on 4 June 1987, and, as the last members of the civil engineering crew made their way out of the tunnel, the Large Electron-Positron installation started up at full throttle.

The LEP's magnets were incredibly heavy, and slotting them into place with a precision equal to a minuscule fraction of a millimetre within tight confines was a challenge and a half. Luckily, the engineers had the assistance of another two dedicated machines to help them. These had a kind of pincer-like feature that earned the robotic pair the affectionate names of 'Lobster' and 'Crayfish'. The effort involved in building the Large Electron-Positron was massive, with vacuum chambers, instrumentation, control systems, water cooling, ventilation and more following the magnets down the tunnel. Everything was tested extensively as the work advanced, with the team constantly on tenterhooks for fear that something might go wrong.

Fast forward to July 1988 and the LEP was approaching a significant moment in its story. Champagne corks were poised, but held back from being released in anticipation as four bunches of positrons shot from the injector to make their 2.5km journey from Point 1 – positioned close to CERN's main site in Meyrin, Switzerland – to the second sector of the LEP's main tunnel, Point 2 of a total of eight, which sat below Sergy in France. The moment was intense, the culmination of weeks spent constructing the machine.

'It worked!' exclaimed Steve Myers, who led the test. 'We learned a lot and it was an extremely useful exercise – exciting and fun to do. Everything behaved as predicted theoretically.' However, despite the euphoria that erupted throughout the control room and was expressed in his enthusiastic report to the collider's Management Board, Myers remained down to earth. He knew the team shouldn't become complacent. 'We [weren't] smug

BELOW View of the Large Electron-Positron tunnel from the cavern at Point 1. *(CERN)*

ABOVE **Physicists gather around a screen in the LEP's control room, anxiously waiting for the machine to be started up for the first time. Director General of CERN Carlo Rubbia stands at the centre, while former Director General of the organisation Herwig Schopper stands to his left.** *(CERN)*

BELOW **Large Hadron Collider project leader Dr Lyn Evans and CERN's Director General Prof Luciano Maiani after the breakthrough of the injection tunnel into the Large Electron-Positron tunnel at Point 2.** *(CERN/Laurent Guiraud)*

because it had worked so well. Crashing testing took four months for about a tenth of the LEP, and at the same rate of testing we required a further 36 months to test the remainder of the machine.'

A full start-up date had been pencilled in for a year's time, and, with limited manpower, everyone worked around the clock on everything – from preparing the software that would seamlessly operate the particle smasher, to leak-testing the vacuum chambers. The LEP's first phase of operation would see it generate electrons and positrons, bringing the particles together in a head-to-head collision. Such a smash-up, according to physicists' calculations, would birth a Z boson with an energy of 90GeV – a neutral carrier of the tantalising weak force and a first step towards determining if the Standard Model was the holy grail of particle physics after all.

Tweaking beams to improve collision rates and testing accelerator components bit by bit, teams of physicists took their positions in the control room and executed 'cold check-outs' – tests made without beams being shot through the collider. They were just days away from running the largest accelerator ever built.

The moment of truth: the great switch-on

Finally, the day had arrived. It was the summer of 1989 and it was time to switch on the Large Electron-Positron for a full run. Physicists called 14 July that year, the day when the collider's first injection took place, an historical moment, a time that we would look back on and remember the very first beam circling the machine. The LEP scientists knew this was to be a 'hot check-out'; particle beams were present for the first time.

That day – and a full 24 hours ahead of schedule – the control room, a small box room behind the SPS control room, was packed. Large audiences filled every corner with bated breath for 50 minutes, the gaze of each and every person fixed on the team coaxing the very first beam around the complete 27km circuit. Then nervousness turned to relief and concentration turned to smiles as the team overcame this first hurdle. But there was still a lot more to be done.

Essential components for the collider continued to be commissioned. The first electron bunches were successfully injected, vacuum pressures were tried and, step by step, the energy at which the particles would be streamed into the main tunnel were ramped up from one electronvolt to the next until the 47.5GeV target was reached. The energy ramp was also followed up by squeezing the betatron – a cyclic particle accelerator that produced oscillations.

Heading into mid-August, the team in the control room watched a screen intently. The physicists had ramped and squeezed electrons and positrons into two separate beams within 32cm. It was late at night and the minutes continued to pass. Gone 23:00 local time, an announcement was made by the physicist behind the OPAL detector. They had just seen the very first collision between an electron and a positron.

The Large Electron-Positron collider was in business.

The LEP's valuable discoveries

Despite their intention to build an experiment as complex as the Large Electron-Positron, particle physicists knew that they would be taking a step into the unknown; a leap of faith that could leave them empty-handed should they have made any miscalculations or overestimated the ability of their newly constructed collider. Now fully built and tested, the team felt they could relax a little. They were ready to take that step into unexplored territory.

Even before the Large Electron-Positron was built, the Standard Model found itself standing on very shaky ground. It was a time when theorists' minds were going exceptionally wild; anything was possible and they wanted to see if the collider could prove their particle-inspired ruminations. To find out, the machine would be operated in two phases. The first phase would run between 1989 and 1995, with almost 18 million Z bosons being made. These particles are short-lived, and the Large Electron-Positron would catch them in the act of decaying into lighter components – a fraction

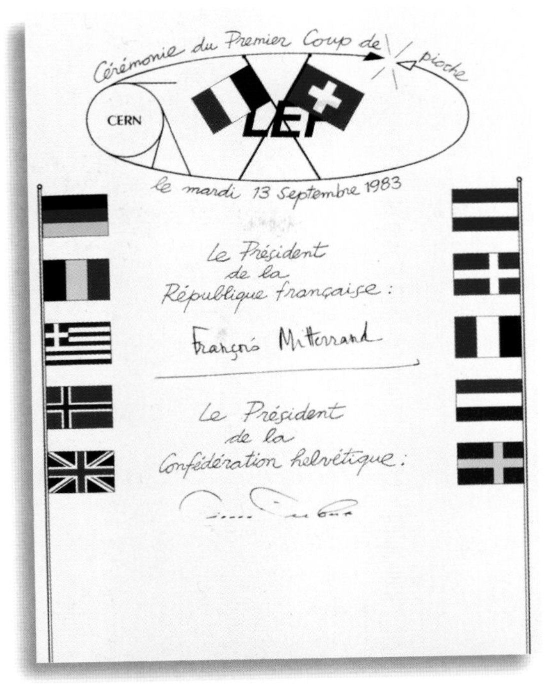

LEFT The parchment that was signed at the Large Electron-Positron Ground-breaking Ceremony. *(CERN)*

of a second snapshot that would prove to be critical in their research.

We know that the Z boson decays in several ways, most commonly in the form of hadrons. However, it also seems to like breaking up into particles that are of the same kind – charged leptons like the muon and anti-muon, the tauon and anti-tauon, for example, as well as the neutrino and its antineutrino. While the decay modes are quite clean, easy to observe events, the neutrino and antineutrino are a challenge; neutrinos and their partner-particles leave the detectors without a trace, so physicists need to look out for such decays indirectly.

As we've already seen, the Large Electron-Positron was beaten to the punch by an earlier collider in finding the Z particle, thus validating the Standard Model somewhat. But there was still much to discover about this weak-force carrier, and the Large Electron-Positron would be the decisive factor in this search. Revving the energies to 100GeV, the detectors – namely DELPHI, ALEPH and L3 – didn't see the Z bosons directly but did see what happened immediately after they were made: the telltale signs of a quark and an antiquark. In the case of DELPHI, physicists witnessed on their computer screens two sprays of particles shooting from the point where electron and positron met in front of the experiment. It's because of the strong force that these newly created particles don't emerge alone. They want other quarks and antiquarks to join them and seize an opportunity to snatch them from the strong force field that surrounds them, hence causing hadron jets such as showers of pions and kaons. Sometimes the violent separation of breaking quarks and antiquarks apart shakes one or two gluons loose, which enable them to fly off to create some trails of their own.

The weeks that followed the beginning of the Large Electron-Positron's first phase would see the collider collect more than 10,000 Z bosons. The general view of this Z particle was becoming multi-dimensional, reaching way beyond our basic understanding of the boson flavour itself. We must remember the conservation of momentum when thinking about the disintegration of particles; the subatomic constituents into which the boson breaks up always weigh less than the Z particle from which they originated. It's here where some groundbreaking discoveries were made, and made sense of the particle lines that criss-crossed the LEP's computer displays.

Take the Z particle's decay into neutrinos, for example. These super-fast decaying particles are often associated with a charged lepton. As the electron and positron beams smashed their way through the collider, the LEP had discovered that only three types of neutrino exist. And given that a charged lepton accompanies every one of them, then there must only be three types of neutrino knocking around the universe too. Given too that there's a symmetry between the quark and lepton families, we'd made an important discovery: there were no more basic building blocks of matter to find. This part of the Standard Model was complete.

When it came to translating the collider's data, the proof was in the resonance peak (see http://backreaction.blogspot.co.uk/2007/12/three-neutrino-families.html). From this it was easy to work out how long the particle lives. It's said that the broader this spike, the shorter its life and the more particles it breaks up into. The breadth of this same peak also reveals how many kinds of neutrinos there are in the universe. Churning out an impressive number of Zs, this particle stood to be the best measured of the fundamental particles, especially thanks to the collider's improved precision over its predecessor. Meanwhile, tweaking the collisional energy between the beams revealed the particle's mass once and for all.

The more Z bosons the accelerator produced, the clearer the relationship between the particles of matter and the fundamental forces became. Finding the tau lepton had already hinted that there were likely to be matching sets of six quarks and an equal number of leptons. What's more, gluons hinted at an unspoken link between the forces. Physicists were about to discover that these are the carriers of the strong force.

After getting up close and personal with the Z particle, the Large Electron-Positron's next task was to uncover more about the W boson. This electrically charged particle, which accompanies the Z boson in carrying the weak force, was, just like its companion boson, discovered before the start-up of the Large Electron-Positron. It had been found in January 1983 by the modified Super Proton Synchrotron, which went by the name of the Proton-Antiproton Collider, with the help of its Underground Area experiments.

Recreating the W particle was actually quite tricky, since they carry an electrical charge. Because of this they don't tend to like being made on their own like the neutral Z boson. Instead, they're made in pairs, one holding a positive charge while the other is negative. They're also heavy, meaning that high energies are needed to make them, and in 1996 positrons and electrons crashed together with an energy of 80GeV.

When it comes to decaying, the W boson has a preference for breaking down into a charged lepton and a neutrino. As we discovered earlier in the case of the Z boson, the neutrinos made can be seen indirectly, zapping away the energy and momentum from the region they once occupied. The particles also don't fly apart when they're made; instead, they tend to form a selection of angles, similar to the formations in which migrating birds fly when they're heading south to escape the winter. When the Large Electron-Positron was notched up to energies above 80GeV, Z and W particles were made together; by cranking it up to around 160GeV physicists could see that production of the 'weak' boson dominated.

Despite the particle taking over the collider, working out its mass wasn't plain sailing by any stretch of the imagination. And physicists had to delve deep into its decays.

W particles don't really have a set way of disintegrating. They break up into whatever form suits them. It's been seen to decay into an electron and a neutrino during a proton-antiproton interaction. Meanwhile in the Large Electron-Positron detector, the W boson broke down into a quark-antiquark pair, while another W particle pairing morphed into a muon alongside a muon-neutrino that managed to evade the detectors, as neutrinos do. The W boson breaks up into all kinds of flavours. That's what made pinning down its mass such a challenge.

However, what CERN scientists had worked out was that the W boson must be incredibly heavy. They knew this because – quite simply – such carriers of the weak force needed to be. It's not felt beyond a minuscule distance of a fermi-metre (that's one quadrillionth of a metre), and it's pretty limited in its range. Where mass and range form a strong relationship, the rules of conservation – which we met earlier – don't apply here, at least not in the micro, quantum world. This essentially means that when the W particle disintegrates it doesn't need to break apart in such a way that energies are precisely maintained. You're probably wondering how this can be. How can one particle break the laws of physics? The answer is that the decay happens so quickly and over such a short range that it's hardly noticed – especially to us in our macroscopic world.

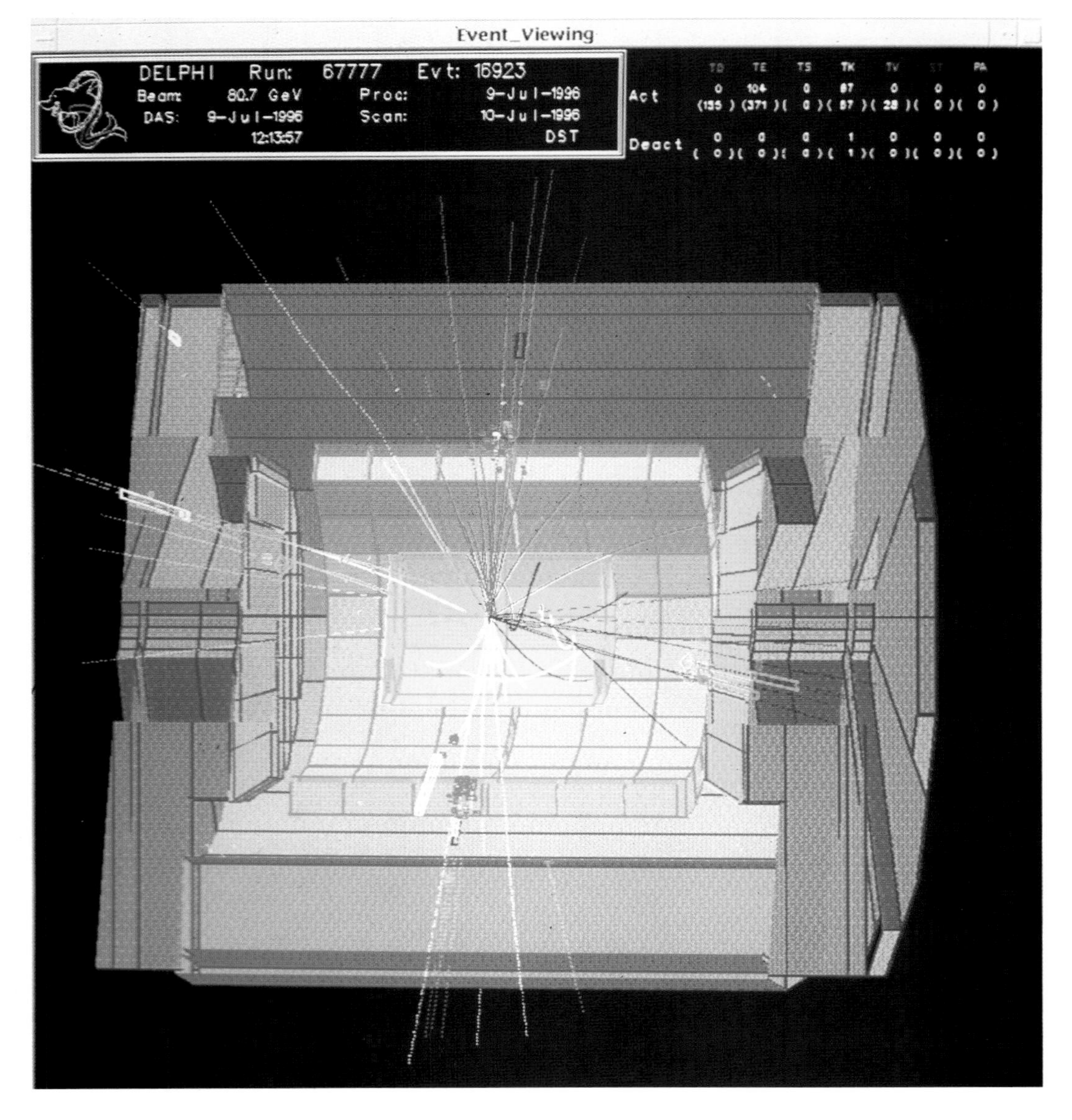

LEFT This is the type of real data that the Large Electron-Positron's DELPHI detector collected when the first pair of W particles were produced. *(CERN)*

ABOVE These low-beta quadrupole magnets could be found either side of the LEP's four interaction points and were responsible for squeezing particle beams to increase collision rates. *(CERN)*

Inside the Large Electron-Positron particles and antiparticles were thrown together, beam to beam, and tended to annihilate each other, a major problem for physicists wanting to create the W particle. In order to get it to create the required signal in front of the detectors, its electric charge needed to be neutralised and by an exact version of itself, yet with an opposite charge, an antiparticle if you will. This means that a positive W particle must always be accompanied by a negative W particle. While it's not easy to create these bosons inside a collider like the Large Electron-Positron, where electrons smash into each other at high velocity, it can be done. As we've seen, physicists just need to turn up the energy.

But the Large Electron-Positron was coming to the end of its run time and, after 11 years of operation, it had come to the end of its tunnel. But physicists had a collider much more powerful in mind to replace it: enter the Large Hadron Collider.

LHC vs SSC: the battle of two great colliders

You could be forgiven for thinking that the transition destined to see the Large Electron-Positron being replaced by the Large Hadron Collider would be a smooth one. But there was a spanner in the works, a problem called the Superconducting Super Collider (SSC). A particle smasher not too dissimilar to the Large Hadron Collider, this particle accelerator would be built in the vicinity of Waxahachie, Texas. Its ring would measure a circumference of around 87km and it would be ramped up to an energy of 20TeV. Without

RIGHT This device was used to inject the antiparticles of electrons – known as positrons – into the Large Electron-Positron ring after being accelerated in the Super Proton Synchrotron (SPS). *(CERN)*

LEFT Inside the tunnel of the Large Electron-Positron, which ran for 27km in a circular corridor between the Jura Mountains and Lake Geneva. *(CERN)*

BELOW A green waterproof membrane can be seen attached to the roof of the Large Electron-Positron tunnel as we look down the corridor at Point 1. *(CERN)*

a shadow of a doubt, the Superconducting Super Collider would be the biggest and most energetic particle accelerator in the world, surpassing the Large Hadron Collider three times over in both size and power.

Despite its strength, being similar to the collider across the pond was causing the SSC problems. Heated arguments erupted between project officials in Texas regarding the Superconducting Super Collider's potential cost, which would be somewhere between an eye-watering $4.4 billion and a heart-stopping $12 billion. Yet construction had already begun, and the process was so well under way that the team behind the collider – predominantly stationed at the University of Texas at Austin – and its supporters, who included Nobel Laureate in Physics Steven Weinberg, had dubbed the yet-to-be-completed machine the Desertron. It felt like there was no going back. Either the Large Hadron Collider and the Superconducting Super Collider would both be built, or the Large Electron-Positron's replacement wouldn't go ahead at all.

Of course, this book wouldn't exist if there hadn't have been some serious consideration that meant CERN's collider got the green light. Sadly for the Superconducting Super Collider, on the other hand, the United States Congress put its foot down, halting the construction of a collider that could have very well possessed the potential to reveal the greatest discoveries of the universe to humanity. The kicker was that, unlike its Texan counterpart, the Large Hadron Collider's tunnel had already been built along with several pieces of infrastructure, lowering the project's cost and meaning that the Franco-Swiss Big Bang Machine could be up and running quickly. By contrast, the Superconducting Super Collider had to be built from scratch, and, with further reports of even higher costs and rumours of poor management by physicists and the Department of Energy officials in charge, the project was officially

BELOW Construction began in 1999 above ground at Point 1 in Meyrin, Switzerland, the location of ATLAS. *(CERN/Laurent Guiraud)*

cancelled on 21 October 1993, not long after its tunnel had been completed and $2 billion had been spent.

The cancellation was a major blow for the United States, who believed that they had lost their leading role in the field of particle physics, with President Bill Clinton attempting to overturn the decision. 'Abandoning the Superconducting Super Collider at this point would signal that the United States is compromising its position of leadership in basic science,' he stated. But it wasn't enough, and 31 October 1993 saw the president signing the bill that would put the plans of the Superconducting Super Collider to bed once and for all. 'A serious loss for science,' Clinton said resolutely. It was game over for the great collider.

CERN makes way for the Large Hadron Collider

The power struggle between the Large Hadron Collider and the Superconducting Super Collider might have been over, but the former still couldn't quite shake off interference brought about by the Large Electron-Positron. The collider had left behind a legacy, standing proud in the memories of its creators as the largest electron-positron accelerator ever built.

Dismantling of the Large Electron-Positron nevertheless began by the end of the year 2000. Just as it was a challenge to put it together, it was just as difficult to break it apart, a tricky business that had to be done with care but quickly, since its successor – the Large Hadron Collider – was waiting in the wings, and engineers were eager to install its components at a moment's notice.

It took 14 months to transport the entirety of the Large Electron-Positron through the narrow shafts that punctuated the tunnel. All in all, engineers calculated it totalled 30,000 tons of material, weighing in at roughly four times the heft of the Eiffel Tower. Some parts of the collider were donated to scientific institutions all over the world, over half of it ended up at the recycling tip, but a significant portion lived on in materials used to strengthen buildings and roads. Meanwhile, boxes of nuts and bolts were checked for radioactivity before exiting the tunnel, with only a small fraction of metal

ABOVE Engineers begin to deactivate the Large Electron-Positron at the end of November 2000 by cutting into the collider's magnets with the help of an hydraulic nibbling machine. *(CERN/Laurent Guiraud)*

BELOW Following the LEP's decommissioning, a superconducting module is lifted out of the tunnel and up to the surface. These modules were features used in the LEP-2 phase, which ran from 1996 through to 2000. *(CERN/Patrice Loïez)*

ABOVE The Large Electron-Positron collider's ALEPH experiment is prepared to be dismantled. *(CERN/Patrice Loïez)*

needing a 'slightly radioactive' sticker applied to their storage containers. Engineers and scientists at CERN were thankful, especially after 11 years' operation. They'd uncovered another vital piece of information: electrons accelerated at breakneck speeds had an incredibly low risk of creating a dangerous, radioactive environment.

That's not the end of the story for the collider, though. One of its magnets is still around

RIGHT Just one of many shafts that give access from the surface down to the Large Hadron Collider's tunnel below. *(CERN/Laurent Guiraud)*

ABOVE The L3 experiment is dismantled at the Large Electron-Positron collider. *(CERN/Laurent Guiraud)*

today, being put to good use as a component of the Big Bang Machine that's taken its place beneath the Franco-Swiss border, filling the same tunnel that the LEP once occupied. The huge octagonal magnet that had once powered the L3 detector was also used for the Large Hadron Collider's detector ALICE, a name derived from its description as 'A Large Ion Collider Experiment'. ALICE is one of the Large Hadron Collider's seven detectors, joining forces with the problem-solving ATLAS, the Compact Muon Solenoid (CMS), the TOTal Elastic and diffractive cross-section Measurement (TOTEM), the Large Hadron Collider beauty (LHCb), the Large Hadron Collider forward (LHCf) and the Monopole and Exotics Detector at the LHC (MoEDAL) to bust the mysteries of the universe, just as L3, DELPHI, ALEPH and OPAL had only months before.

Soon to be left with an empty tunnel where the Large Electron-Positron had buzzed with particle-acceleration activity, engineers had another task ahead of them; at that moment in time, there was nowhere to put the seven detectors particle physicists had set their hearts on.

From unearthing an ancient relic to underground water: digging out detector caverns

The excavations that had taken place in the 1980s to accommodate the Large Electron-Positron had left CERN with a beautifully crafted tunnel, four caverns, and several access shafts that led off from the main corridor. Engineers reasoned that the smallest of the Large Hadron Collider's seven experiments, ALICE and LHCb, would fit snugly into a pre-sculpted cubbyhole. However, the more sizeable CMS and ATLAS detectors would need large rooms, caverns that would need to be excavated further. And there was competition between the teams behind these two great detectors – they were both keen on having Point 1, which was handily situated a stone's

throw from CERN's main campus and far away from Point 5, the most remote location in the tunnel. With Points 2, 4, 6 and 8 still housing the Large Electron-Positron, there wasn't much choice. Point 3 was deep under the Jura Mountains and incredibly inhospitable, while Point 7 would have been an entirely unrealistic location for the two detectors. Placing their hard hats on, a brand new team of engineers had to begin tunnelling again.

It was the type of rock that resolved the argument over location between the CMS and ATLAS teams. During an extensive underground investigation between 1995 and 1997, engineers boring holes into horizontal layers of sedimentary rock at Point 1 noted a mixture of sandstones, marls and several transitional rock types. These ground conditions weren't perfect, but they were suitable enough for ATLAS. Consequently CMS would have to settle for Point 5.

'Funnily enough, Point 5 was one of the worst places possible to build the cavern in terms of geology, though best from a physics point of view,' recalls project manager for CMS civil engineering John Osborne, amused. 'And CMS was pretty much the only experiment that could cope with being built here.' On digging

ABOVE Just one of the access shafts used during construction, which led into the Large Electron-Positron collider's tunnel. They are still in use today for access to the Large Hadron Collider. *(CERN)*

RIGHT Experimental halls are excavated to enlarge the Large Hadron Collider (LHC)'s experimental caverns where its detectors would later be housed. *(CERN/Laurent Guiraud)*

ABOVE An aerial view of Point 1, where the ATLAS detector is located 100m underground. *(CERN/Laurent Guiraud)*

LEFT The Compact Muon Solenoid (CMS)'s cavern at Point 5. *(CERN/Maximilien Brice)*

ABOVE Boreholes were drilled along a proposed line of the Large Electron-Positron tunnel under the Jura. This is to find out what type of conditions are likely to be encountered during the collider's construction. *(CERN)*

further into the cavern at Point 5, engineers got more than they bargained for. Pushing deeper and deeper into sandy marls and marly molasse of varying density, they came across unconsolidated debris that would have struggled to support ATLAS. They also came across the foundations of some kind of ancient relic. 'The first thing we found was the last thing you would ever want in a construction site,' laughed the Large Hadron Collider's project leader Lyn Evans, recalling the day they made the great find. They had to stop digging.

Six months elapsed as archaeologists replaced engineers and toothbrushes took precedence over earth-movers at the CMS cavern. Excavation continued, but at a much slower and more careful pace. Dusting away dirt and stone slowly but surely revealed what had temporarily halted further progress on the Large Hadron Collider. It was a Roman farm, a Gallo-Roman villa with surrounding fields; jewellery, medals and silverware were scattered in the vicinity. Coins, which were discovered to be minted in London, Lyon and the nearby ancient harbour city of Ostia, were dug up and dated all the way back to the fourth century. 'This proves the United Kingdom, at least during the fourth century, was part of a single European currency,' jokes Evans.

It turned out that engineers and archaeologists alike were standing on a very important site in the eastern French town of Cessy. Go back to 121 BC and they stood in what had been the war zone where Julius Caesar and the Celts clashed to decide the mastery of Gaul. Julius Caesar and his Legions had won, of course, with the Battle of Alesia ending in a Roman victory and that led to the expansion of Roman power over the entirety of Gaul.

RIGHT The empty cavern that would one day be home to the ATLAS experiment. *(CERN/Laurent Guiraud)*

LEFT During the excavation of the cavern for the Compact Muon Solenoid, engineers unearthed a Roman villa dating back to the fourth century. *(CERN)*

It turned out that the farm villa that had been discovered lay to the side of the detector shaft, enabling the Large Hadron Collider engineers to carry on with the excavation of the CMS' cavern even as the archaeologists worked. Teams from two completely different professions and excavation styles worked tirelessly side by side, toiling away for months on end. When it was time for lunch, CERN's team would watch the Gallo-Roman villa seem to slowly but surely rise out of the dirt; as the CMS engineers ate their sandwiches ancient pots, coins, stone walls and even graves appeared before their very eyes.

The decision was eventually made to remove some of the farm's stone walls and transport them to a museum in France. Unfortunately for both teams, though, the farm's boundaries overlapped the area where the CMS detector would be set up. Consequently there was

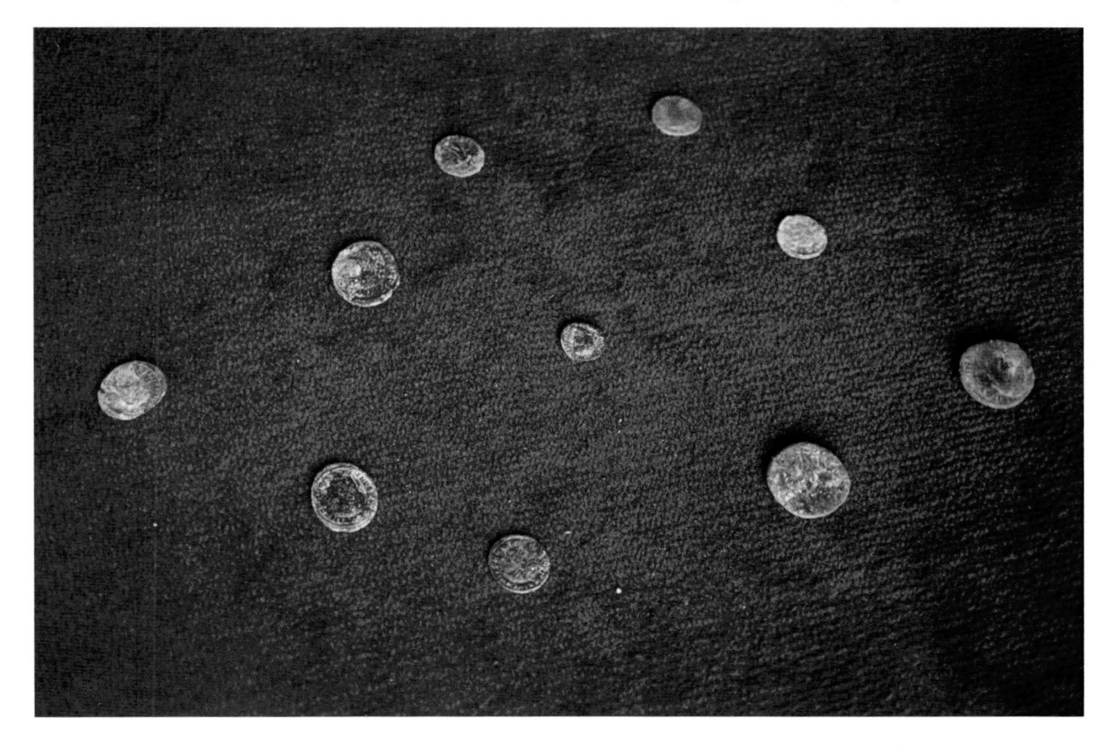

LEFT Roman coins dating back to the fourth century, excavated from the site of the CMS cavern. *(CERN)*

ABOVE As well as construction of experimental caverns, buildings were built above ground to store the Compact Muon Solenoid (CMS) detector as well as other components before they are lowered into the pit. *(CERN/Laurent Guiraud)*

only one thing for it. 'They are buried again,' explains Osborne, 'but they are protected so archaeologists can return to excavate them again in the future.'

After carefully covering the ruins with layers of blankets, rock, gravel and sand, the tunnel diggers could finally continue with their excavation of the Point 5 cavern. Previous borings had revealed another snag, however. Ground water flowed underfoot and it wasn't until the engineers sank shafts into the floor that they realised how quickly the liquid seemed to race past rock, stone and other pieces of debris. To get around the problem, the team had to freeze the ground, doing so by pumping brine into the pre-bored holes, cooled to some -25°C (-13°F) in a bid to turn the liquid into a 3m-thick wall of solid ice. But the water racing from Cessy still came thick and fast, and the two shafts that had been forced into the ground made matters worse by creating a channelling effect between them and building up so much pressure that the water had no choice but to penetrate the walls.

The engineers paused, standing around and scratching their heads. They needed a much more heavy-duty liquid – something that could freeze water in a matter of seconds. They went for liquid nitrogen, plummeting the water temperature to -95°C (-139°F). It did the trick, halting the water and creating a wall of ice around the shafts that was solid enough for them to continue digging. As ever, though – and even during the days of first excavating the tunnel for the LEP it seemed the engineers could never catch a break – there was another issue at Point 5, and the team were forced to stop and regroup again.

Ironically it was the same reason why ATLAS couldn't occupy Point 5 in the first place: the ground was too soft. 'There was only

RIGHT A view of the Compact Muon Solenoid (CMS)'s experimental cavern, during its construction in the year 2000. *(CERN/Laurent Guiraud)*

about 15m of solid rock and for the first 75m of digging down it was just a type of glacial deposit called moraine, a mixture of sand and gravel,' says Osborne. 'The rock we did have was a kind of soft sandstone called molasse. A large cavern built in this would just collapse.'

Refusing to submit to this new spanner in their plans, there was only one thing for it – the engineers had to build some kind of large, supporting structure under the surface; something able to cope with the hefty mass of soil layered over it whilst simultaneously holding up the caverns where the detectors would be nestled. A concrete pillar seemed to fit the bill nicely, serving as a divider between two great experimental halls. It turns out that this pillar would also have another use. 'Knowing that we would have to build this structure anyway, we asked radiation teams how thick we'd need it to be to protect people from radiation in the cavern next door to the experiment. They gave a figure of 7m. The width needed to ensure adequate support for the caverns was 7.2m, so this worked out very well,' says Osborne. 'The second cavern could also be safely used when the machine is on.'

These lulls in progress had put the project back somewhat, forcing the engineering teams to look at ways of clearing away rock at a faster rate. Going back to when the Large Electron-Positron tunnel was being excavated, explosives were on everyone's minds, a method of blowing away sandstone, dirt and limestone in mere minutes. For the CMS cavern, though, sticks of dynamite didn't seem to be doing the trick, which left the team to resort to Plan B: they would excavate the cavern little by little before quickly splattering the walls with concrete. The engineers called this process 'shotcrete', and in less time than it took to be ready and poised with their cement spray guns the fine, liquid sheet of concrete would hit the walls, drying as quickly as it made contact. Steel anchors bored some 12m deep in the surrounding rock finished the job, leaving the rest of the team to move on to the next section. This was how the excavators worked for some time, a sequence of excavating, blasting with shotcrete and anchoring with steel before moving further along the cavern and starting all over again. 'If we didn't do this, the whole thing would have collapsed whilst we were building it,' explains Osborne. 'We constantly monitored any movement with a host of instruments and adjusted the support as necessary.'

In the areas that they couldn't reach, workers used a lift fashioned like a cage that was lowered up and down the access shaft, manipulated with ease by pulleys and ropes. The system was ideal, allowing workers on site to reach the required wall level easily and quickly. However, despite the number of times the workers used it to get the job done it did have a daunting aspect. 'Being on a rope meant that it had a tendency to sway as you went down,' laughs Osborne, 'and 100m is a long way!'

And the workers couldn't forget the water that had splashed up through the ground months before. They had frozen it, but there was every chance that it would thaw, returning to a body of gushing liquid and turning soft rock into a slippery mud, incapable of supporting the 12,500-tonne CMS detector. Waterproofing, drainage systems and painting the cavern were a must; the longevity of the Large Hadron Collider was hugely dependent on it. 'We sealed off the cavern with waterproofing and put in a permanent concrete wall up to four metres thick, reinforced with steel bars,' explains Osborne. He and his team were also mindful

BELOW Cranes can be seen, ready to lower personnel and equipment down into the pits below. *(CERN/Laurent Guiraud)*

of the environment up on the surface, as they toiled away day after day for several years. All in all, almost 250,000m^3 of soil and rock was ripped out from Point 5, and shifting so much debris threatened to cause a great deal of noise and road disruption. To avoid aggravating residents of the area this way, the engineers had a trick up their sleeve.

They figured that depositing unwanted molasse and moraine close to CERN's main buildings was the ideal solution. You'd be forgiven for thinking this would create an eyesore, but the CMS team immediately began layering it with fresh soil, before planting it with vegetation. That bump in the landscape is still visible at CERN today, an artificial hill that serves as a reminder of the challenges, hardships and adversity that the engineers faced in laying the foundations of just one component that would contribute to the world's greatest Big Bang machine.

At Point 1 workers discovered seemingly better conditions than in Point 5's cavern, but they still remained cautious, aware of one of the many challenges that stared back at them as they looked down the length of the tunnel – not just the fickle nature of rock they had encountered, but also because another particle smasher was already happily in operation, smashing particles and antiparticles together while its detectors looked on: because at this time the Large Electron-Positron wasn't due to be dismantled until early November 2000. So with the collider still up and running at the time of excavation, the digging out of the hall where ATLAS would sit took a more unconventional route.

BELOW **The ATLAS cavern begins to take shape, in this photograph from September 2003.** *(CERN/Maximilien Brice/ Patrice Loïez)*

Ask any civil engineer how they usually sculpt out an underground cavern and they'll tell you that you need to tunnel downwards, down to where you'll lay the floor, before building the walls up and finishing with the ceiling. But since the engineers didn't want to disturb the Large Electron-Positron – which was running straight through the cavern – they decided to cast the ceiling first. Some 10,000 tonnes of concrete and steel would go into a 2m-thick vertical wall, held in place by an impressive 38 cables bolted into the shafts.

Reaching this point, the team had to wait until the Large Electron-Positron had been powered down before they could dig the tunnel any further. When that day came the newly completed ceiling already hung over them as they pulled out 30,000 tonnes of rubble before laying the floor and erecting the walls. Once the ceiling was supported the cables supporting it could finally be removed. Such was the precision in excavating CMS' chamber that the ceiling barely moved when the workers severed the final cable. Dusting their hands together, the team took a step back to marvel at their work. The cavern was complete.

Despite the precision in excavating Point 1, engineers knew that the cavern would actually move. Quite a worrying image might have filled your imagination on reading this. It sounds unbelievable, but this alcove had pushed upwards, against 60m of solid rock, much like a bubble of air forces its way to the top of a glass of fizzy cola – but with a 7,000-tonne detector nested inside it. The amount of rock pulled out of Point 1 weighed in at a lot more, far greater than the weight of ATLAS – which was the bubble in this case – despite its heft. This 'anomaly' in the rock was causing it to rise towards the surface. 'This was predicted by the civil engineers,' explains Peter Jenni, a former spokesperson for the ATLAS experiment. The engineers estimated a rate of rise of about 0.2mm per year, and they were ready and

prepared; every piece of equipment was placed in the cavern with this shift in mind, while the floor of the hall, which wound up being 5m thick and reinforced with a mixture of concrete and steel, joined the movement – essentially without warping. If this movement hadn't been allowed for, then the entire project would have been on a path to failure.

When it was completed, ATLAS' two caverns were impressive. They were arranged at a right

ABOVE Digging machines come to rest, signalling the end of the excavation of the Large Hadron Collider's tunnel. *(CERN/Maximilien Brice)*

LEFT The excavation site of ATLAS' detector. *(CERN)*

RIGHT **Civil engineers sculpt the cavern where the ATLAS detector will sit; one of seven experiments of the Large Hadron Collider unravelling the secrets of the universe.** *(CERN)*

angle, with the main cavern being 53m long by 30m wide. Its roof hung 35m above the yet to be filled hall. The service cavern by its side was designed to be smaller, with a diameter of 20m and a length of 65m, connected to the detector's gigantic chamber by five L-shaped tunnels, measuring 2.2m to 3.8m across. Walk along these tunnels and you'll discover two other caverns. It's in these rooms that you'll find the electrical equipment that gives the LHC machine life.

So that's two of the detectors housed inside the tunnel, and we know that the ALICE experiment would take the place of the L3 detector, reusing the magnet that once powered it; but what of the four other experiments? At this moment in time the teams behind them had yet to convince CERN that they were essential to the particle collider's mission.

CERN's observing partners

While excavation continued below the frontier between France and Switzerland, conversations, handshakes and a flurry of important documents were exchanged behind the scenes. Japan had expressed interest in becoming an observer state of the great physics experiment. Headed by Kaoru Yosano, its Minister of Monbusho (Education, Science and Culture), Japan was the first country from outside Europe to be accepted into CERN with Official Observer status. The honour would allow his country to attend council meetings and receive documents that would outline the future of the Large Hadron Collider.

As Mr Yosano and his country were welcomed into the fold, applauded by representatives of CERN's then-19 Member States* on that June day in Geneva in 1995, the organisation had another reason to celebrate as grins and warm congratulations filled the room: Japan were keen to see further improvements in the construction of the Large Hadron Collider and, aware of the hardships and setbacks that had been faced in constructing the tunnel, wanted to lend a helping hand that would speed up the process. A momentary silence filled the room in CERN's headquarters: the Japanese government had offered the particle physics organisation financial aid to the tune of 5 billion yen, around 68 million Swiss francs (£50 million).

Closing the meeting, CERN's then-Director General Professor Christopher Llewellyn Smith thanked Mr Yosano and accepted a Daruma doll from the Japanese Minister. This was a Japanese cultural tradition, the doll's one completely painted eye marking the beginning of a partnership. Its second eye wouldn't be drawn

*Bulgaria, Israel and Romania had yet to join.

LEFT The ATLAS cavern had to be anchored from above, since there would be no room for support pillars within the cavern once the experiment was built. *(CERN/Laurent Guiraud)*

BELOW A huge tunnel was built from the surface down to the ATLAS cavern through which the giant segments of the experiment could be lowered. *(CERN/Laurent Guiraud)*

in until the final component of the Large Hadron Collider had been bolted in place. It was a huge turning point for the experiment. 'I am convinced that Japan's newly acquired Observer Status along with its contribution to the LHC Project will serve to further strengthen the cooperative linkage between the Japanese and European scientific communities,' said Mr Yosano to the room. 'I am looking forward to this project opening up new frontiers in the field of particle physics and advancing joint international efforts to pioneer these new domains.'

It would be two years later, around Christmas 1997, that the United States would follow in Japan's footsteps. They too wanted a stake in the Large Hadron Collider, drumming up $531 million with the intention of also becoming an Observer State. Eagerly signing the agreement, directors of the Office of Energy Research, the National Science Foundation, put pen to paper as members of CERN looked on. The country would contribute to the ATLAS and CMS detectors we met earlier and, as we shall see later in the chapter, it was the start of an ever-lasting relationship between the two organisations. '[Together we'll] extend the achievements of physics, so stunning in the 20th century, into the 21st. The Large Hadron Collider will be the largest [step] so far in the march to global scientific collaboration,' said Martha Krebs, Director of the Office of Energy Research, during her speech. 'It will give testimony to our conviction that the wonder and joy of discovery about our world and the universe are suitable activities for humankind even as we will always struggle with poverty and violence. It will inspire young people of all nations that the quest of science will bring benefits to their world. I am honoured to be here with you today.'

Filling the Large Hadron Collider's caverns

The teams behind the TOTEM, LHCb, LHCf and MoEDAL detectors fought hard for their place in the design of the Large Hadron Collider. Construction of the great particle smasher was well under way at this point. The LHCb detector would be the fourth experiment to be approved by CERN, followed quite swiftly by the other three. After all, time was of the essence and physicists had their eye on a start-up date in September 2008. They had to get constructing if they were to meet that target.

Back at Point 1, it was 4 June 2003 and a ceremony was taking place. It was a special occasion, after all. After a lengthy five years of digging, cementing and ingenious civil engineering, ATLAS' cavern was fully excavated.

RIGHT Inside the surface buildings at Point 1, where the A Toroidal LHC ApparatuS (ATLAS) experiment is housed. *(CERN)*

Looking around the hall as a musician blew into the mouthpiece of an alphorn, Pascal Couchepin – the president of the Swiss Confederation – was joined by officials of the experiment, CERN and other members of Switzerland's political authority. As high as a six-storey building and twice as long, you could fit the United Kingdom's Canterbury Cathedral quite snugly inside the cavern where they stood. After five years, the first hall was complete and ATLAS could be installed.

Out of the seven detectors at the Large Hadron Collider, ATLAS is the general-purpose one, alongside its sister experiment, the CMS. It's the machine that dabbles in all kinds of different physics, particularly the kind you'll find at incredibly high energies – conditions that are likely to have existed when the universe was nothing more than a mere cosmological infant, when temperatures were so fierce that particles would have continually popped in and out of existence.

ABOVE A view of the A Toroidal LHC ApparatuS (ATLAS) experiment in 2006. *(CERN/Laurent Guiraud)*

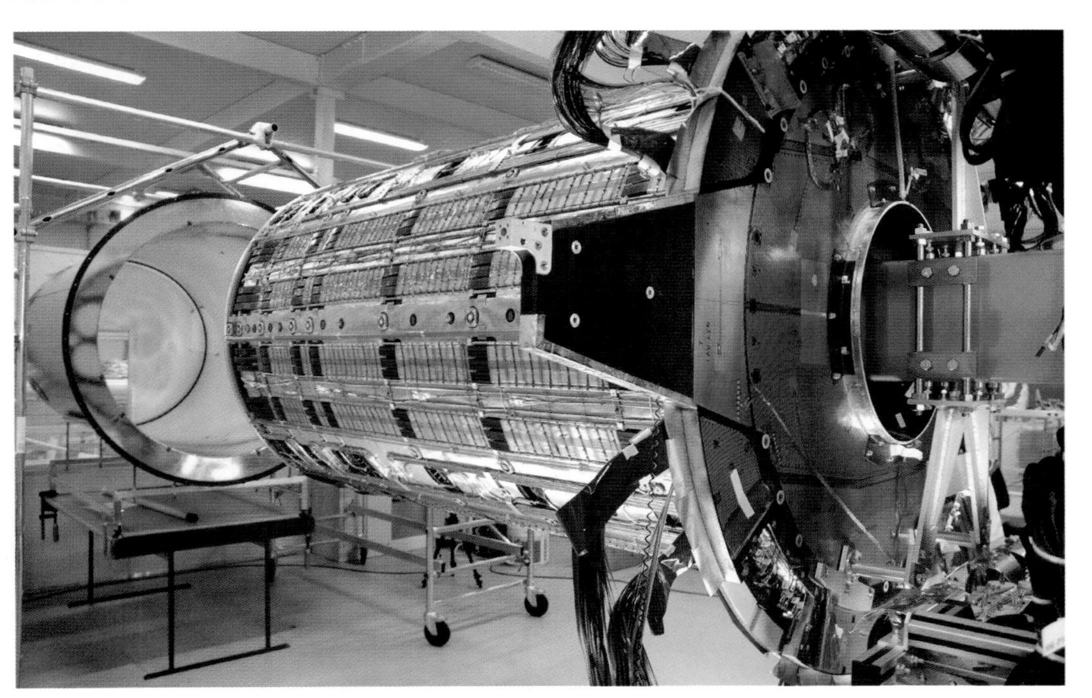

LEFT A component of the Large Hadron Collider's ATLAS detector is constructed in a clean room. Here the silicon microstrip detector system is slotted into the experiment's Transition Radiation Tracker (TRT). *(CERN/Serge Bellegarde)*

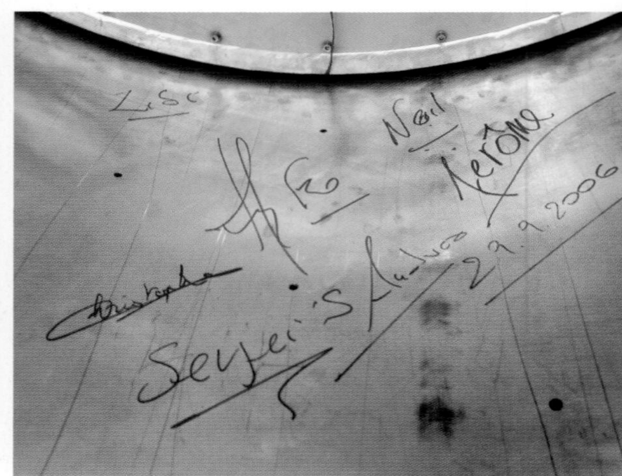

ABOVE The first of ATLAS' small wheels is lowered into the cavern. *(CERN/Claudia Marcelloni)*

ABOVE RIGHT Before the subdetector is assembled completely, the ATLAS team leave their mark on the experiment. *(CERN/Serge Bellegarde)*

RIGHT An ATLAS magnet end-cap is transported to Point 1, where the experiment is installed. *(CERN/Claudia Marcelloni)*

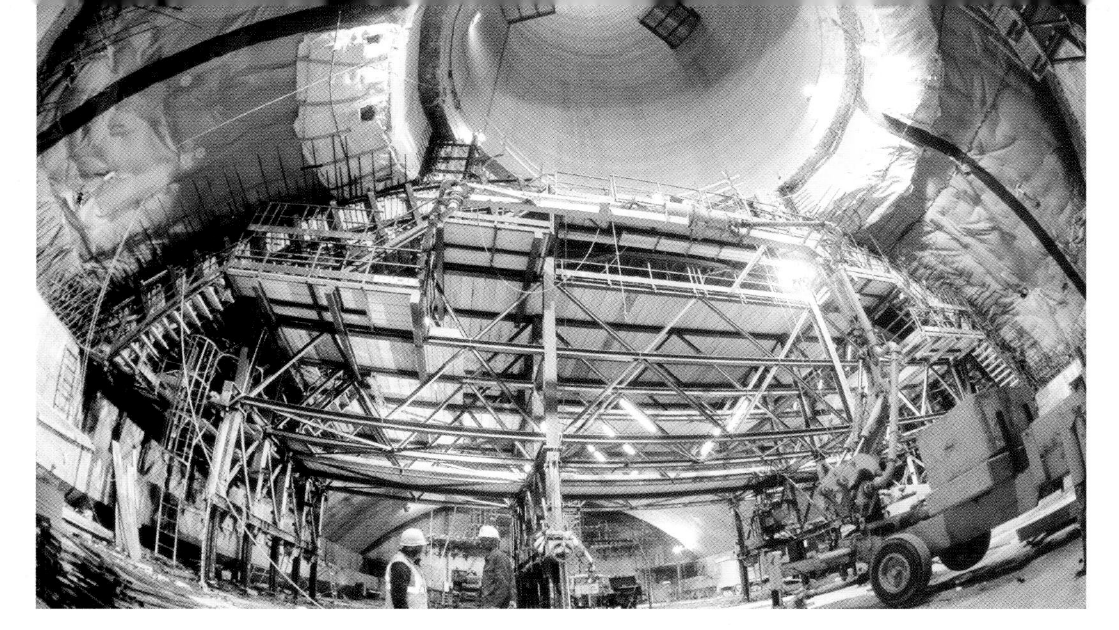

LEFT The framework for ATLAS takes shape in this photograph taken in June 2001. *(CERN/Laurent Guiraud)*

Once constructed within the Large Hadron Collider, it was ATLAS' job to sift through these particles. What the machine seeks to determine is, what are the basic building blocks of matter? What really are the fundamental forces of Nature, and could there be some kind of great, underlying symmetry to our cosmos? ATLAS is an explorer, testing everything from the predictions of the Standard Model to the possibility of extra dimensions and what could make up the ingredients of dark matter. Its aim is to leave no stone unturned in our quest to make sense of our cosmos.

If you stand in front of ATLAS, or have had the pleasure of doing so already, you'll understand how the detector steals the show. Its aluminium supports, cables, chambers, trackers, liquid helium pipes and superconducting magnets are beautifully entwined, hence its impressive size that fills a

BELOW Assembling ATLAS' calorimeter cryostat, which was filled with liquid argon. *(CERN/Maximilien Brice)*

ABOVE
1. **The first stage of lifting the ATLAS end-cap upright.**
2. **A crane begins to lift the ATLAS end-cap.**
3. **The ATLAS end-cap weighs 160 tonnes.**
4. **The upright ATLAS end-cap.**
(CERN/Maximilien Brice)

RIGHT **The huge coil for the first ATLAS toroid magnet to be installed is delivered upon the back of a very long truck.**
(CERN/Maximilien Brice)

LEFT An unusual wide-load: an ATLAS toroid magnet coil. *(CERN/Maximilien Brice)*

BELOW The 1,380 square metre ATLAS cavern has to be so big to harbour the 46 x 25 x 25 cubic metre experiment, which is the largest of its kind in the world. *(CERN/Laurent Guiraud)*

ABOVE ATLAS half-completed, with six of the toroidal magnets assembled. *(CERN/Maximilien Brice)*

OPPOSITE Engineers connecting millions of wires to ATLAS' electromagnetic calorimeter that is affixed to the instrument's end-cap. *(CERN)*

BELOW The Compact Muon Solenoid (CMS) being built at Point 5. *(CERN/Valeriane Duvivier)*

35m by 55m by 40m cavern. Indeed, ATLAS is a mechanical giant, a factor that became all too apparent during its construction underground.

'The words "ATLAS installation" don't nearly do justice to the magnitude of the task,' says Robert Eisenstein of the Santa Fe Institute, a member of the detector's team. 'During a three-and-a-half-year period, from April 2003 until December 2006, more than 7,000 tonnes of large, delicate apparatus was lowered into UXI5 [the technical name for ATLAS' cavern].'

If you had wandered along the Large Hadron Collider's tunnel to Point 5 between November 2006 and January 2008 you would have seen CMS taking shape. Just like ATLAS, it's a detector that has a broad programme of intricate particle physics to investigate. On paper, the pair have exactly the same scientific goals: CMS watches for the same new physics phenomena that the Large Hadron Collider might turn up. In essence it isn't too different from a camera. There's probably been many a time where you've pressed the shutter of your DSLR or the photography application on your smartphone in order to get that perfect motion shot. The Compact Muon

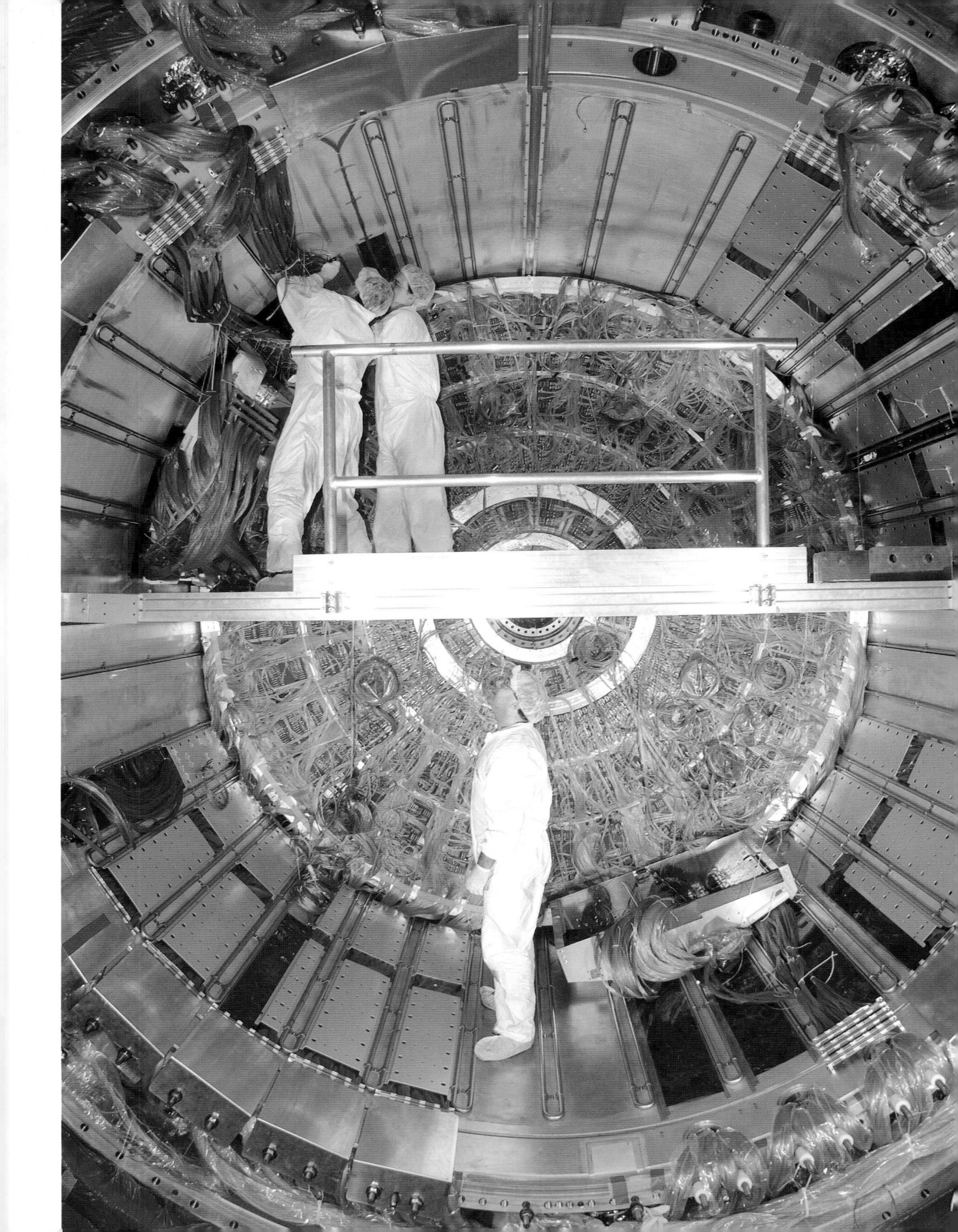

ABOVE A beautiful mountainous backdrop was the scenery for engineers building the cavern for the CMS experiment. *(CERN)*

LEFT In November 2005, the Compact Muon Solenoid (CMS) Tracker support tube arrived in the new Tracker Integration Facility (TIF) at CERN Meyrin. *(CERN)*

BELOW An elevator was attached to a crane to allow engineers to descend into the CMS cavern as they built the experiment. *(CERN)*

RIGHT In 2005, the final piece of the Compact Muon Solenoid (CMS) magnet arrives, kick-starting a ceremony held in the CMS assembly hall at Cessy, France. *(CERN)*

Solenoid uses a similar kind of technology, except that it snaps three-dimensional photographs of particle smash-ups – as soon as they occur – at up to 40 million times per second.

Despite its impressive speed, the CMS, as we've discovered, is one hefty machine – the weightiest of the Large Hadron Collider's entire suite of detectors. This meant that transporting it down a gaping hole in the Swiss landscape was never going to be taken lightly. Engineers sliced the detector up, opting to transport it down the shaft before setting it up 100m below ground. 'It was a huge amount of responsibility,' says Hubert Gerwig, who was responsible for ensuring that each component arrived safely. 'You have all the physicists waiting for an element to arrive, and if you make a serious mistake it could have been the end of CMS.'

It wasn't an entirely new method, though. The team had been inspired by the LEP, which had undergone a similar experience in which 300- and 350-tonne components for the L3 detector were built on the surface before being fed down to ground zero. Inside the cavern, CMS was built in a Russian doll-style, each piece remaining accessible while engineers worked in parallel as they navigated challenging cable chain loops, water, gas and cooling leads. They were building the great detector from the inside out. Needless to say, using this method meant they had to be very careful. 'As soon as I saw the sketch, I had a clear vision of how such an experiment had to be organised,' explains Alain Hervé, who was CMS' original technical coordinator. 'We were already two years inside Large Electron-Positron running, and lessons from the construction and the first shutdowns were clear to me. The way the experiment had to be sectioned and installed was the first priority, not developing the design.'

The excavation of the cavern, the pits and surface hall were all designed with the 'lowering method' in mind. As soon as engineers started scooping out the Point 5 alcove, they had

The large disc of the CMS detector, which weighs in at 400 tonnes, is slowly and carefully lowered 100m beneath the Earth's surface. *(CERN/Maximilien Brice/ Claudia Marcelloni)*

ABOVE The enormous CMS cavern, which is 53 x 27 x 24 metres. The elevator car attached to a crane that descends 100m below the surface to deliver workmen into the cavern can be seen in the background. *(CERN/Laurent Guiraud)*

ABOVE RIGHT The CMS cavern completed and awaiting delivery of the experiment itself through the overhead tunnel in 2005. *(CERN/Maximilien Brice)*

RIGHT The giant solenoid that creates the magnetic field ten thousand times stronger than the Earth's magnetic field in the CMS experiment. Here we see it stored in an assembly hall above the CMS cavern, awaiting installation. *(CERN/Maximilien Brice/Adrian Billet)*

their endpoint mapped out; it was all part of a carefully formulated plan. A massive, strong 'plug' was built into the design, especially for holding pieces of the CMS before lowering them down the shaft. Cranes then took the strain, whilst anchor points were used for a spot of heavy lifting. As the last piece of the detector disappeared slowly down the shaft, touching down on the subterranean surface of the cavern, Hubert and his team breathed a sigh of relief. 'Everything had been calculated and calculated,' he explains. 'We were successful and nothing was damaged, so there was a sense of relief. It was a bit like an exam – you feel better once it's over and you can celebrate!'

RIGHT Assembling the CMS' end-caps. *(CERN/Laurent Guiraud)*

RIGHT The ALICE experiment's subdetector known as the Dijet Calorimeter (DCAL) is installed, along with a further upgrade to ALICE's other subdetectors. *(CERN/Arora Harsh)*

At Point 2 another substantial amount of excavation had been taking place. Engineers were keen to slot the ALICE detector into the concrete painted room, not too far from ATLAS. Of all the Large Hadron Collider's major experiments, this one was known as the 'small detector', despite weighing in at 10,000 tonnes and measuring some 26 x 16 x 16m. At 56m below France's St Genis-Pouilly, it's quite a specific experiment. Its task is to unpick the whys, whos and wheres of strongly interacting matter, particularly where energy density is extreme and where conditions are such that quark-gluon plasma feels right at home. This is an incredibly hot, dense soup – including a cocktail of all kinds of exotic particles – like that which is thought to have existed shortly after the Big Bang, a mere few millionths of a second after the birth of the universe. It's here that ALICE can get a clearer picture of what happened when this plasma expanded and cooled before giving rise to the particles that so abundantly compose the matter we see around us today, the basic building blocks of cosmic make-up.

BELOW The A Large Ion Collider Experiment (ALICE) Time Projection Chamber during assembly in 2007. *(CERN/Antonio Saba)*

Look at the ALICE experiment and you'll have a feeling that you've seen it somewhere before. That's because you have. It's taken on the role of being chief carer of the L3 detector's former – and incredibly characteristic – huge, octagonal, fire-engine red magnet, four storeys tall and set slap bang in the middle of the cavern that once belonged to the LEP's former detector. Its engineers had a challenge beyond that of excavation, for that had been completed after the very first particle smasher had been given the go ahead. What they had to do now was find a way to get 30 chunks of a gigantic, multimillion-dollar dipole magnet through a narrow passageway, without clanking into the magnet already positioned 20m below. They didn't want to disassemble L3's monstrous magnet, which could have led to damage

RIGHT The last of the Large Hadron Collider's dipole magnets is lowered into the tunnel. *(CERN/Maximilien Brice)*

as well as cost additional time and expense. Instead, they intended to carry each of the 33-tonne pieces of the brand new muon dipole over the 'hurdle' and slip it down the other side. It was almost like a game of Tetris; it required a great deal of concentration to get to the very end of the game, a game that took a good several months and required a gut-wrenching amount of skill and care.

There was no denying that it was an incredibly tight squeeze. Engineers had to walk a tightrope of manoeuvring gigantic and heavy pieces of equipment whilst at the same time ensuring the pieces didn't swing wildly beyond the confines of 2cm to 3cm. If they got it wrong, it would have spelt disaster for the detector. Imagine the set-up of your house relying solely on being able to push a huge piece of furniture through a narrow doorway, then magnify that anxiety tenfold and spare a thought for the man who was in charge of installing ALICE, project leader Detlef Swoboda. After lowering all 30 chunks down the shaft and getting them over L3's solenoid at a painstakingly slow few centimetres per minute with the help of a crane, he and his team then had to assemble the new magnet, test it, then break it apart again in the small confines of ALICE's new home. Despite the intensity of the project, though, Swoboda took it in his stride. 'I'm a designer and a builder of magnets,' he says modestly. 'It was just my job.'

Assembly of the Large Hadron Collider was very much at full tilt. And, within schedule, ALICE, ATLAS and CMS were almost at the end of their completion processes. Magnets swayed and cables flexed as the engineers worked steadily towards the particle smasher's switch-on. In terms of detectors, there was now just one major experiment to find a home for: the LHCb.

RIGHT The Large Hadron Collider's ALICE detector is installed piece by piece and placed into the cradle where it studies the collisions of beams of lead ions. *(CERN/Maximilien Brice/Claudia Marcelloni)*

There's a slight difference between matter and antimatter and the 5,600-tonne LHCb wants to know why. To some degree it's keen to know what ALICE understands; what conditions were like straight after the birth of our universe, when the two types of particle are thought to have dominated in equal measure. The LHCb is a robotic detective, going back to the scene of the crime and trying to figure out why antimatter was given the boot, leaving galaxies, stars and even us to be dominated entirely by matter.

LHCb is therefore seeking out the beauty quark, or bottom particle, and its antiparticles. It's a pretty elusive member of the Standard Model that prefers not to stick around for long, disintegrating within a trillionth of a second into lighter particles. That's why the experiment is designed the way it is – devoid of a barrel-shape and comprising a series of subdetectors sitting behind each other, stacked uniformly like books in a library or bricks in a wall. Which is how it doesn't miss a thing, each of its subdetectors always watching and specialising in seeking out a variety of characteristics – such as energy and momentum – about the B particles (which are composed of the bottom or beauty quark) that spray out from protons in a head-on collision. 'B particles are very unstable and decay only a millimetre away from the collision that produced them,' LHCb physicist Richard Jacobsson explains. 'LHCb has to be able to confirm whether it's detecting particles that come from the collision, or from a "secondary vertex" a short distance away, which could indicate the presence of a B particle.'

Ask any of the engineers where they were looking to put LHCb and they would have known straight away. They had their eye on the cavern where the Large Electron-Positron's DELPHI had once stood; Point 8, not too far from Ferney-Voltaire in France, on the border of Switzerland. Large enough? Check. Already been excavated? Check. Unlikely to accrue further cost to the project? Check. The pit behind Geneva airport looked like the perfect place to station a particle-hunting detector.

Two years after switch-off of the Large Electron-Positron, engineers were champing at the bit to get the final major piece of the Large Hadron Collider's design into place before they could see the light at the end of the tunnel, figuratively speaking. LHCb's cavern isn't too

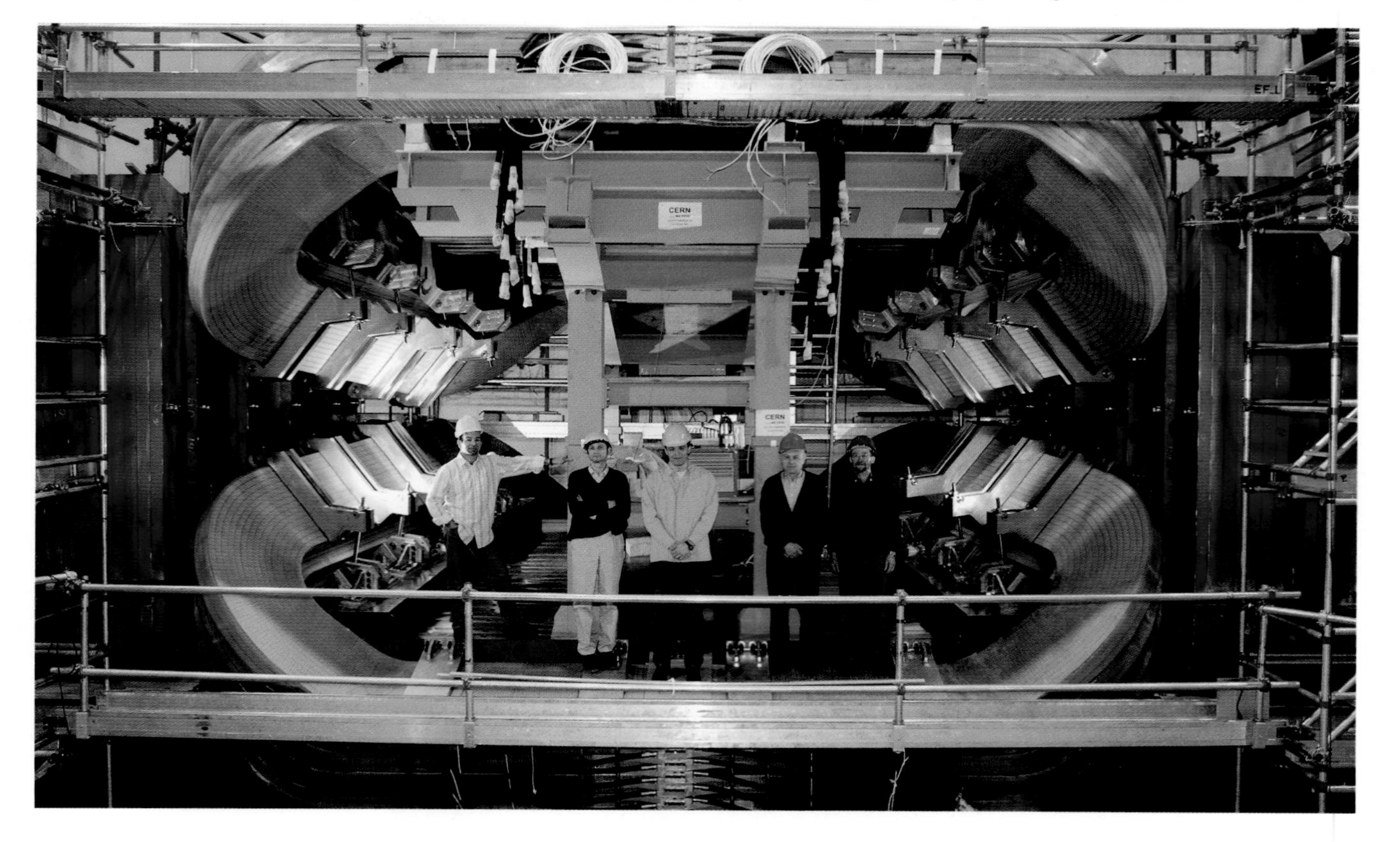

BELOW The LHCb detector (pictured with the experiment's team members) is assisted by magnets in order to help the LHC investigate the difference in the amount of matter compared to antimatter. *(CERN/Maximilien Brice)*

dissimilar to the ones that house ATLAS and CMS, just smaller and lined with grouted steel rock bolts 12m long. The fastenings are topped off by lashings of fibre-reinforced concrete, interlaced with steel mesh that's up to 1.3m thick. The hall was ready for use, but that didn't mean that the hard work was over.

What do you do with a 1,600-tonne magnet that's too big for the narrow shaft you want to pass it through? If you're a CERN engineer installing the LHCb, you'll break it apart and reassemble it on the other side. And that's exactly what happened with the experiment's biggest and heaviest component. 'It was built underground piece by piece,' recalls Rolf Lindner, who was in charge of the detector's installation. 'It was definitely the most difficult to install – we estimated it would take three months to assemble, but it ended up taking almost a year.'

LHCb's detector – RICH-2 – was less trouble, but it was one of the experiment's most fragile components. It was therefore packed together on the grounds of CERN before being bundled up on a lorry to be driven 8km – at a speed akin to that of a tortoise – and guided down the chute to Pit 8. Aware of the flow of traffic near the site, engineers took to moving RICH-2 by night. Cars, buses and lorries were diverted away from the site, plunging the area into darkness and silence, save for the gurgle of the vehicle transporting the precious cargo and the steady LEDs and lamps that silhouetted the engineers at work. At a maximum speed of just 1kph, the transport vehicle struggled under the strain. As its axles twisted and turned, the tubes of steel split under the weight, as effortlessly as snapping a twig that's fallen from a tree. This happened twice, so that the vehicle didn't reach the opening of Point 8 for another 24 hours. Then RICH-2 was promptly – but carefully – slotted into place.

Installing a Large Hadron Collider is a tricky business, and dangerous to say the least. The team behind the LHCb experiment knew this only too well as they handled fragile pipes laden with beryllium, a highly toxic substance when airborne. Lindner remembers this process as if it were yesterday. So breakable were these tubes, which needed to be threaded through the detector, that a screwdriver dropped from a few metres above would have shattered them. Pushing the final piece of pipe into the LHCb during the summer of 2007, engineers were relieved to say that any potential crisis had been averted. At least for now.

ABOVE The Monopole and Exotics Detector (MoEDAL) uses detectors to search for a particle called the magnetic monopole. *(CERN/Maximilien Brice)*

The LHCb also needed a protective wall. Erecting this was a precarious task, with concrete blocks required to seal off the detector entirely. 'The blocks had to fit very tightly together in order to protect against radiation,' Lindner explains. 'We couldn't afford to leave even the smallest of gaps.' Indeed, if any had been left radiation would have seeped from the LHCb experiment during operation, exposing technicians in the control rooms to waves of harmful emissions, whilst proton beams crashed and smashed into one another recreating what conditions were like just after the Big Bang. The wall was designed to be several metres thick, and 7.5 tonnes of concrete slabs needed to be piled up on a wonky cavern floor. Engineers had to get everything just right. Fortunately a helping hand was available in the form of a crane, which expertly lowered the blocks down the shaft before piling them on the floor to be slid across the cavern.

The LHCb has a companion at Point 8, and that's MoEDAL, the primary interest of which is to search for an hypothetical particle with both electric and magnetic charges – a magnetic monopole. Imagine a magnet that has a north pole but no south, and vice versa. The magnetic monopole is kind of like an isolated magnet, preferring to have a net charge rather than two sides of opposite polarity; in short, it's exactly what its name says – a magnet with one pole. If you cut a magnet in half, and even if you carried on chopping it up, you'd still have the usual north and south arrangement. A monopole, by contrast, would only have one magnetic pole, and when it came into contact with a magnetic field it wouldn't follow the usual north–south alignment; instead it would race off into infinity. Physicists have another name for these particles, especially when they have a zero electric charge: dyons. Generally speaking dyons supposedly exist in four dimensions, exhibiting both electric and magnetic charges. If monopoles are out there somewhere, then they are most likely exceedingly massive, making their presence known by their magnetic charge and their incredible ionisation powers that would see them have a ripping force supposedly 4,700 times stronger than the proton.

They are just two types of particle that MoEDAL likes to hunt, but its quarry extends to other stable, massive particles that like to ionise, ripping electrons from the matter they crash into as well as pseudo-stable massive particles. Being so ionising, particle physicists know when they're around, since they have a habit of damaging anything with which they come into contact. But not MoEDAL, which has nuclear track detectors built so solidly and impenetrably – featuring a plastic top layer and an aluminium bottom sheet – that they're perfectly suited to such aggressive particles, along with other such radiation. Spotting the furiously etched-in tracks that give away a newly found particle's mass, charge, energy and the direction in which they're flying, is a piece of cake for this experiment, despite its simplistic design. 'Our cost is at least a factor of 100 cheaper than the other Large Hadron Collider,' explains James Pinfold, MoEDAL's spokesperson. 'If we did discover the monopole it would be a real David versus Goliath configuration.'

MoEDAL is a detector and camera rolled into one. It's only really sensitive to new kinds of physics, snapping and trapping anything groundbreaking that happens to come within its line of sight. For this reason, the CERN scientists like to call MoEDAL the 'passive' experiment, which only really sits up and takes notice when the particle it's looking for decides to show up.

Physicists have been hunting for the monopole for quite some time, even during the days of the LEP. And they're really hoping it exists. It would solve a whole deluge of problems in physics, from the untidiness of Maxwell's equations, which underpin all the electric, optical and radio technologies you see around you today, to being the piece of puzzle that would tie together the electromagnetic with the weak and strong nebula forces at high energies. In fact, theoretical physicists think there should actually be loads of monopoles just waiting to be found.

But we could be wrong about what we're looking for. There's a possibility that monopoles don't behave or have the form that physicists have envisaged. It could be that they're much heavier than we thought, or we may never be able to detect them. 'If the monopole existed in those too-massive ranges, we'd never be able to produce enough energy on Earth to create them,' explains Richard Soluk, MoEDAL's technical coordinator based at the University of Alberta. 'Let's hope they're not in those too-heavy-to-create ranges.'

ATLAS might be a big experiment, filling up a good proportion of its cavern, but that doesn't mean it's unable to share its living quarters. Nestled quite closely to its network of magnets, subsystems, metal and wiring, you'll also find the LHCf. Made up of two detectors, this experiment is kind of like ATLAS' assistant, with both of its sensors sitting some 140m either side of all the action. That's the point where proton beams collide. LHCf serves almost like an extra pair of eyes, whilst the larger ATLAS intently watches the exact same interaction point. As you've probably gathered, LHCf is actually quite small. In fact it's the tiniest of the Large Hadron Collider's experiments.

Its job on the Large Hadron Collider is to make 'pretend' cosmic rays. These streams of high-energy particles actually come naturally from outer space, bombarding our planet's upper atmosphere, crashing into the atoms and causing a cascade of pions that are so unstable they disintegrate into a selection of particles including photons, protons, antiprotons, electrons and positrons. It's here where the

experiment can get stuck into studying these neutral particles before they rapidly fall apart, pinning down where ultra-high-energy cosmic rays actually come from – something we've been trying to work out ever since astronomers John Linsley and Livio Scarsi spotted one shooting across the array of detectors composing the Volcano Ranch experiment in New Mexico back in 1962.

That's not the last of the more pint-sized members of the Large Hadron Collider. There's also TOTEM, which is similar to LHCf in one respect; it too serves as an extra experiment for a much larger detector, but in this case the CMS in the Point 5 cavern. Keeping an intense look-out at its interaction point using a couple of 'telescopes', T1 and T2, TOTEM is interested in the lowly proton, how it behaves, and its characteristics as the subatomic particles crash head-to-head right in front of its sophisticated, silicon detector 'eyes'. There are two kinds of collision that interest TOTEM: what are known as inelastic collisions, where one proton survives while the other produces 'lighter particle' debris and ploughs forward; and elastic particle smashes, where both protons make it out alive and are sent flying in different directions and angles.

TOTEM does the things that the other detectors at the Large Hadron Collider can't. It's able to take measurements of particles that scatter at tiny angles, watching out for the particles that brush past each other rather than crash head-on, thanks to its workforce of three detector types – Roman Pots packed with microstrip silicon detectors that pick up on the protons, Cathode Strip Chambers, and Gas Electron Multiplier (GEM) detectors that have the knack of catching the jets of particles that pop out of the subatomic crashes when the quarks of which they're composed fly apart. TOTEM is also CERN's longest experiment, with its detectors spread over half a kilometre around the CMS interaction point. It's the machine intended to fill in our knowledge of Quantum Chromodynamics – the theory that addresses the strong interaction where quarks and gluons are glued together.

You'll find this strong interaction between the hadrons or family of quarks that make up

ABOVE The TOTal cross section, Elastic scattering and dissociation Measurement (TOTEM) at the Large Hadron Collider. *(CERN/Maximilien Brice)*

the proton, neutron and pions – especially in searingly hot, thick, dense places like the very early universe. As has already been explained, the ALICE experiment dabbles in such conditions, as does CMS, but it's TOTEM that's interested in how the size and actual shape of a proton varies during energetic collisions – especially since CERN had previously discovered that protons are much more likely to collide when energies are at an all-time high. This detector is looking into the phenomena much deeper, giving the other Large Hadron Collider experiments a heads-up in calibrating their luminosity monitors, ready for the particle-smashing show to begin.

BELOW The Large Hadron Collider's superconducting magnets are tested in the assembly hall. *(CERN/Laurent Guiraud)*

We'll delve deeper into how each of these detectors works in the next chapter. Meanwhile there's still work to be done in the main tunnel before powering up the great particle accelerator.

Installing the super-magnets

If you fancied a 27km walk, 100m underground – a circular stroll that begins at CERN in Meyrin and takes you close to the Jura Mountains, continuing underneath the lush French countryside right through to Switzerland's Geneva airport and then back to CERN again – then you'd notice a common theme along your route: the eight sections of cylindrical tubes that alternate between blue and silver for every kilometre you progress. Inside these tubes, a lot of tech lives, serving as the driving force behind the particles in the collider: the superconducting magnets. These highways, which speed particle beams in opposite directions, are cooled to a frigid -271°C (-456°F), which is theoretically the lowest temperature that can be attained close to absolute zero and much colder than outer space. If you were keen to count these magnets during your walk, you'd get to a total of 5,000 – an impressive quantity for what would be an impressive machine.

During a March Monday lunchtime in 2005, civil engineers laid aside their sandwiches. They had something much more important on their minds: a giant superconducting magnet, some 15m long and weighing in at 35 tonnes, was being lowered down a shaft, while a specially designed vehicle waited below ready to take it to its final resting place. Just one of the many jigsaw pieces that would join up the Large Hadron Collider's experiments, this piece signalled that the end of construction was in sight. The engineers would install roughly 20 magnets every week hereafter, aligning them precisely before welding them tightly together.

As they carefully continued to shuffle magnets down the tunnel, the legacy of the LEP seemed almost to live on in the minds of the engineers navigating the underground network. By comparison, the former particle smasher's magnets were smaller and much lighter. The clue to the difference in scale is, of course, in the name *Large* Hadron Collider. The fragile

RIGHT **Short, straight sections containing magnets for manipulating the beam inside cryostats – of which contain liquid helium – are assembled for the Large Hadron Collider (LHC).** *(CERN/Maximilien Brice)*

BELOW **Due to the amount of magnets and cryostats that engineers were faced with, they often called their working area 'Lego Land'.** *(CERN/Maximilien Brice)*

BELOW RIGHT **Director General Luciano Maiani unveils an electronic information panel that indicates the number of Large Hadron Collider's dipoles still to be delivered, coupled with the days left until the deadline of 30 June 2006.** *(CERN/Patrice Loïez)*

LEFT **Each short, straight section is cooled to -271°C (-456°F) to ensure optimum performance.** *(CERN/Maximilien Brice)*

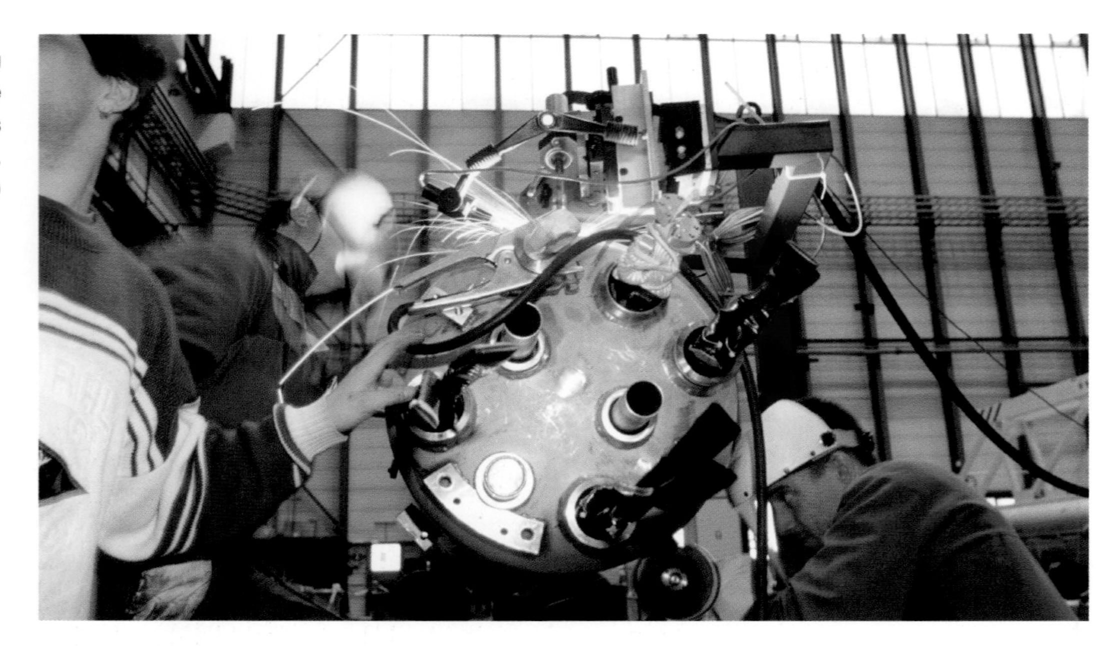

RIGHT The welding of one of the Large Hadron Collider's dipole magnets. *(CERN/Laurent Guiraud)*

tube of magnets was a tight squeeze in such a narrow passage, whence an overhead crane had carefully lowered them and through which transfer tables helped them on their way.

Around two years after the very first magnet touched down at ground zero, the very last was slotted into place. It was late March 2007, and the engineers were about to run tests, eager to see if the magnets were up to the job. More than 1,200 magnets had been packed inside the 12,500-tonne detector, in which 392 of them would keep the particle beams on track.

Waiting in the tunnel, the engineers were startled by a gigantic explosion that ricocheted off the walls and forced them to clap their hands over their ears and flee the vicinity as fast as they could. A quadrupole magnet, that would be essential in focusing the collider's proton beams, had ruptured. As engineers from CERN and Fermilab pored over their design documents, both laboratories were keen to speak to the public. 'Documentation for the parts were reviewed four times without an error being spotted,' said Judy Jackson, CERN's director of communications. 'We're going to be asking how this happened.'

It was Fermilab, who had designed and built

LEFT The Large Hadron Collider's superconducting magnet test string located in the assembly hall. *(CERN/Laurent Guiraud)*

the magnet before its installation, who held up their hands to take the blame. The test-cycle that had been initiated was to observe how the magnet would respond to an increase in temperature way beyond the normal range expected when the Large Hadron Collider was switched on. This sort of increased temperature causes these types of magnet to leave their superconductor state behind, forcing them to become resistant to electrical current, and the freezing liquid helium surrounding the magnet consequently begins to boil off, exerting enormous amounts of pressure on the entire structure. But the 'inner triplet' magnets struggled to cope with the resultant 20 atmospheres of pressure, and began to shuffle uncomfortably. The helium swirling around them released an impressive amount of energy, ramping up the pressure along one of the three magnets to a metal-crushing intensity, and a support made of a fibreglass-epoxy laminate had snapped.

'What the analysis showed was that something extraordinarily simple was missed in the design,' explains Pier Oddone, Fermilab's director. 'Even though every magnet was thoroughly tested individually, they were never tested with the exact configuration that they would have when installed at CERN – thus missing the opportunity to discover the problem sooner.'

However, though the engineers had hit a snag they weren't going to let that get in the way of firing up the Large Hadron Collider. 'The goal is to get it [fixed] so it doesn't affect the start-up schedule,' Judy Jackson said confidently. 'We don't see anything that tells us we can't do that.' And she was right. The particle collider was on schedule ... but only just. It had already been delayed by two years. This was second-time lucky.

Unsurprisingly, negative rumours began to circulate amongst physicists, with whispers of construction delays at the world's largest particle accelerator continually being exchanged. It was now October 2007, and CERN still had their eye on a 2008 debut, but a maelstrom of blog updates on the collider exaggerated the problems that had been encountered, forcing CERN's Director General Robert Aymar to step forward. 'There have been no show stoppers,' he insisted. 'We can all look forward to the Large Hadron Collider producing its first physics in 2008.'

Testing... testing...

Such was Aymar's confidence that CERN had announced the day they would be switching on the Big Bang machine: 10 September 2008. Meanwhile, behind the scenes, engineers were eager to run a series of tests. The Large Hadron Collider would be the world's most powerful particle accelerator. It would be able to make beams of particles some seven times more energetic than any machine ever built before. It also had the capability of creating conditions 30 times more intense, maximising the limit of its design performance. The Large Hadron Collider was, in some respects, its own prototype, being powered by technologies that wouldn't have been possible mere decades before it was built.

It was consequently the very first of its kind, and, when just moments remained before it was due to be started up again, engineers remained nervous. Spanners tightened, cranes lowered components and trucks drove engineers to underground sites to put the finishing touches to the great machine. By this time it was July 2008, and all eight sectors of the collider had begun the long process of cooling down to 1.9°K, just above absolute zero. All 1,600 superconducting magnets went through a series of electrical tests, both alone and working together as one unit, to ensure they were all functioning as they should.

You could be forgiven for thinking that turning on the Large Hadron Collider is as easy as flicking a switch. It isn't. It's a long process that focuses on each and every part of the collider's components as they get up to speed. The next stage is to ensure that the Large Hadron Collider and the Super Proton Synchrotron (SPS) accelerator – which injects a single bunch of particles into the former's ring – were synchronised. This procedure forms the final link in the injector chain, and timing is everything. The pair have to communicate – with the SPS ready to release and the Large Hadron Collider ready to accept – within a fraction of a nanosecond, a mere one-thousand-millionth of

ABOVE Workers concrete the walls of the vault that connect the Super Proton Synchrotron to its injector. *(CERN)*

a second. 'The test couldn't have gone better,' says the collider's project director, Lyn Evans. And sure enough, the protons worked their way along the expected 3km, right to the end of Sector 12, on the very first attempt. During this trial the particles picked their way through the collider's network of magnets in a clockwise direction. Physicists would need to fire protons in an anticlockwise direction for the true test. That was planned for a few weekends later.

'We opened the tap and it went straight through,' says Paul Collier of CERN's operations group, almost breezily. It was true that scientists experienced a few issues with their software and timings, but, overcoming them effortlessly, the team were stunned by something else in the stuffy control room in which they stood: 'The quality of the machine alignment was superb,' adds Collier. The corrector magnets didn't need to be adjusted – not even by mere millimetres – to steer the protons around.

Waiting in the wings was head of the accelerator and beam department Steve Myers. He was on the lookout for problems – something with which he had become all too familiar during his work on the LEP, when he and his team of engineers had come across two beer bottles stuffed into the Large Hadron Collider's predecessor's pipes. So Myers was alert to any blockages in the Large Hadron Collider's pipes, but the successful tests alleviated his worries. If there was a problem, they would know about it. 'There are in the order of 2,000 magnetic circuits in the machine,' he said. 'This means there are 2,000 power supplies that generate the current which flows in the coils of the magnets.'

Just earlier that day, during a test of the collider's beam-injection system, the physicists had popped their champagne corks after watching particle debris erupt from protons and smack into a block of concrete. It was as if time had stood still. No one spoke, as all eyes remained fixed on the action unfolding on their screens. It marked the moment when, for the very first time, particle tracks had been reconstructed from a man-made event generated by a collider.

The test was designed to put the LHCb experiment through its paces. Waiting for the green light from the CERN Control Centre, the atmosphere was tense. The Super Positron Synchrotron spat a few billion protons down the 2.7km transfer line into the 28-tonne slab of concrete, the collimator that waited at the entrance of the Large Hadron Collider, some 200m from the Point 8 cavern where the detector sat. 'It was quite overwhelming, actually,' said Themis Bowcock, who led the team.

But Bowcock need not have worried. The particle pile-up shot straight through the block, racing along the Large Hadron Collider's pipes to create hits of electricity in the selection of silicon discs that adorned the Vertex Locator (VELO), the instrument that has the ability to track the particles generated by the proton-proton collisions and sits just a few millimetres away from the action. The control room erupted in applause and cheers. The test had gone much better than anyone could have ever hoped. 'It was amazing to have seen the first Large Hadron Collider tracks,' grins Bowcock.

The team raced through tests that weekend, taking note of the beam and producing further tracks in the LHCb. They'd successfully tested an eighth of the particle collider with a beam of protons.

The big switch-on

It was 10 September 2008 and the doors to particle physics that had been securely locked since the dawn of recorded time were about to be flung wide open. It was the big day. The Large Hadron Collider was about to be switched on, a project that had taken years to complete, cost billions to build and had been plagued by construction problems, cost overruns and equipment issues. But as teams of engineers, scientists and the world's press stood in the control room, all that seemed like a distant memory.

But there was one remaining concern. The general public were afraid that the great collider would destroy the Earth by generating a devastating black hole, a region of space where gravity is so strong that light – indeed, nothing – can escape it. Worried phone calls and angry emails flooded CERN as it struggled against the backlash. 'We received a lot of worried calls from people,' particle physicist and Large Hadron Collider spokesman James Gillies remembers. 'There was nothing to worry about. The [collider] is absolutely safe, because we've observed Nature doing the same things the Large Hadron Collider will do. For example, protons regularly collide in the Earth's upper atmosphere without creating black holes.'

Nevertheless, CERN felt it had no choice but to release a report dispelling the notion that particle accelerators were dangerous. The Large Hadron Collider had a group of experts to investigate such possibilities, and they'd done their homework long before construction of the collider was complete. They were the LHC Assessment Group.

In essence, the report that was issued echoed Gillies' statement and revealed that if the particle smash-ups that would occur inside the collider had the power to tear our planet apart, then we wouldn't even have existed in the first place; the regular crashes between Earth's atmosphere and cosmic rays would have already obliterated our planet as well as other bodies throughout the universe. Indeed, the cosmic rays that rain down on us from space are more energetic than the Large Hadron Collider could ever be – the collider's exotic particles wouldn't race around with as much speed. In fact, even if cosmic rays were capable of creating black holes, they'd already be sitting on our planet right now, gobbling up its matter at an alarming rate. 'Each collision of a pair of protons in the [collider] will release an amount of energy comparable to that of two colliding mosquitoes, so any black hole produced would be much smaller than those known to astrophysicists.'

Back in the control room the clock was ticking and the world was watching. It was almost 10:30 in Geneva. All eyes were fixed on the screens that adorned the walls and the computer screens that lined the curved desks throughout the room. A beam of protons made their way around the 27km ring for the very first time, successfully steered with precision as each section of the collider opened in turn. At each of the points where ALICE, CMS, LHCb and ATLAS stood, the beam was halted as collimators – the devices that narrow the beam – snapped shut, the idea being that corrections to the particles' flow could be corrected if necessary. If everything was running as it should, these barriers would allow the particles on their way to proceed through the detectors further along the tunnel.

On the screens in front of them, the experiments lit up in turn. The sequence worked like clockwork; particles raced to collimators, collimators swung open, beams raced to the next collimators, and so on. The atmosphere was electric. They'd done it. 'There it is,' Lyn Evans said as the beam finished its lap and as two white dots flashed up on the coloured screen. His statement was met with euphoric cheers.

A second anticlockwise beam fired around the ring just a few hours later, taking slightly longer than the hour it took for the first beam to complete a lap. There was apparently a problem with the collider's cryogenics, slowing progression from injection at Point 8 through to Point 6. It took the beam 30 minutes to reach Point 5, where the CMS experiment was nestled. But this minor problem didn't dampen any spirits – after all, the second beam had made 300 turns of the collider. 'The machine worked beautifully,' said Evans.

Throughout operation the proton beams were to be steered in opposite directions in the Large Hadron Collider, racing at a breakneck speed close to the velocity of light, completing some 11,000 laps every second. Evans and his team might have made a major step in circulating the first beam, but they knew that there was still more to be done. The real work for the operations team was just beginning. They needed to actually snatch the particles for study. That was one of the fundamental purposes of a particle accelerator.

For that, they needed to switch on the radio frequency, or RF system. This is a metallic chamber packed with an electromagnetic field, to supercharge the particles and accelerate them with its electric fields, which keep the particles in their packets of billions of protons. If these forces didn't exist the particles would just end up flying apart as they're forced through the tunnel thousands of times per second. Evans had worked on the conversion of the Super Proton Synchrotron when CERN decided to transform it into a proton-antiproton collider, and he was worried. If past experience was anything to go by, noise – particularly from the klystron, a vacuum tube used to amplify high radio frequencies – had a habit of splitting particles apart, causing them to fly off in all directions.

Fortunately, the very first test on 11 September revealed that Evans needn't have worried. The beams had created a mountain-range plot – a graph line that reveals a series of peaks, akin to the plot's namesake – confirming that the RF had worked its magic. If it hadn't, then the particles would have strayed from their perfect orbit around the machine. What onlookers would have seen instead would have been the 'mountain range' broadening at a rapid rate before slumping and flattening out. 'It was the real champagne moment,' says Evans. The Large Hadron Collider had made a perfect profile, creating a long and continuous narrow ridge, providing the team with a view of what the particle bunches were doing inside the accelerator: they were passing the same point at the right time, without fail.

The Large Hadron Collider's meltdown

But Evans and his team were to be faced by their first failure after the switch-on. It came just a few days later on 19 September 2008, while engineers were happily running through powering tests of the main dipole circuit in Sector 3–4. Physicists had been passing 8,000A of electricity through this portion of the underground ring when a tonne of liquid helium began to seep into the experiment's circular tunnel. There was nothing else for it, the collider had to be shut down, at least for the time being, whilst physicists tried to figure out what had gone wrong. 'Coming immediately after the very successful start of the Large Hadron Collider's operation, this was undoubtedly a psychological blow,' said CERN's Director General Robert Aymar. 'We'd had better days,' added James Gillies.

But what had happened?

The collider is designed to race protons around like Grand Prix cars on a circuit, at energies of 7TeV, before crashing them together like bumper cars at a fairground. The beams not only have to be fast, they also have to be incredibly focused. Enter a total of 9,600 superconducting magnets designed for such a job, made of a mixture of metallic niobium-titanium. The problem was that these components struggled to fulfil the task. They'd been carrying huge loads of electricity, and when they began to overheat, sweltering and sizzling above their operating temperature of -271°C, or -456°F, their helium liquid coolant began to boil off and the Large Hadron Collider ruptured with an almighty boom.

What remained at the joint, where the electrical connection laced the accelerator's magnets together, was like a scene from a car crash; the magnets looked like they'd smashed into each other and melted in such a way that it looked like there had been a hugely violent explosion. After a rapid increase in temperature to 100°C (212°F) some 100 magnets had lost their superconductive properties in a blip called a quench.

It was just one of tens of thousands of joins around the collider's circuit, but it was enough to keep the accelerator offline for months, blowing any chance of a restart before the winter maintenance period – when shutdown was important to save on electricity costs – out of the water. 'Events occur from time to time that temporarily stop operations, for shorter or longer periods, especially during the early phases,' said Peter Limon, a physicist at CERN.

'It wasn't such a big deal,' offered high-energy physicist Mike Harrison, who is based at Brookhaven National Laboratories in New York. 'The actual fix took a day or two. The problem [with particle accelerators] is that you have to warm them up and cool them down again. That's what takes up the time.'

It turned out that, while engineers and physicists could have done without the catastrophe, they were prepared for it, and had all the necessary procedures in place. 'It does seem that all the systems that are supposed to protect the machine in cases like this worked,' said Gillies.

The collider might have been out of action, but its inauguration still took place. Japanese Vice Minister of Education, Culture, Sports, Science and Technology T. Yamauchi was ready, poised with a marker pen that would signal the project's completion in traditional Japanese style: he drew in the second eye of the Daruma doll.

ABOVE In March 2014, engineers achieved a major goal in the installation of the Large Hadron Collider's DFBA splices. *(CERN/Anna Pantelia)*

The long shutdown

The collider didn't accelerate any particle beams until the following year. Even hours after the quench, physicists still had their hearts set on their very first particle beam collision sooner rather than later. But first there was work to be done. Relief valves were drilled, a quench-protection system was installed to shut the accelerator off at abnormal voltages, magnets were heated to enormously high temperatures and cooled back down again, and engineers picked their way along the tunnel hunting for electrical faults between the collider's magnets that would have sent them back to square one.

All in all, a total of 53 magnet units were pulled out of the Large Hadron Collider's system to be cleaned or repaired. Some needed to be taken to the surface, before being cast aside and replaced. By June 2009 the collider was supercold, ready for more powering tests to bring it back to life – and with a whopping electricity bill that added roughly $10 million to the accelerator's costs. 'We built this machine to operate it,' said Steve Myers. 'If you buy a Rolls-Royce, you can afford to put petrol in!' The CERN physicists were not going to be put off.

Just over a year after the quench the physicists were back within the control room's sleek blue confines once more, cheering and slapping each other's backs for a job well done as the first beam circulated. It was 20 November 2009 and the collider was powering up again, but this time at half the energy it was designed to attain: 7TeV. The CERN physicists weren't taking any chances. This time they were going to take it slow and steady.

At this point the Large Hadron Collider was far better understood than it had been when it broke down a year earlier, said Myers. 'We'd learned from our experience, and engineered the technology that allowed us to move on. That's how progress is made.'

At roughly 22:00 that evening a proton beam made its way around the Large Hadron Collider. The machine was back in business, ready to delve deep into solving the mysteries of the universe. There was just a touch of ramping up to be done first. 'It was great to see the beam circulating in the Large Hadron Collider again,' said CERN Director General Rolf Heuer. 'We still had a way to go before the physics could begin, but with this milestone, we were well on our way.'

Chapter Three

How a particle smasher works

A particle's journey through the Large Hadron Collider is one of incredible speeds, being squeezed along a network of magnets to create groundbreaking physics.

OPPOSITE A 3D cutaway of one of the dipole magnets that propel protons and ions around the LHC.
(CERN/Daniel Dominguez)

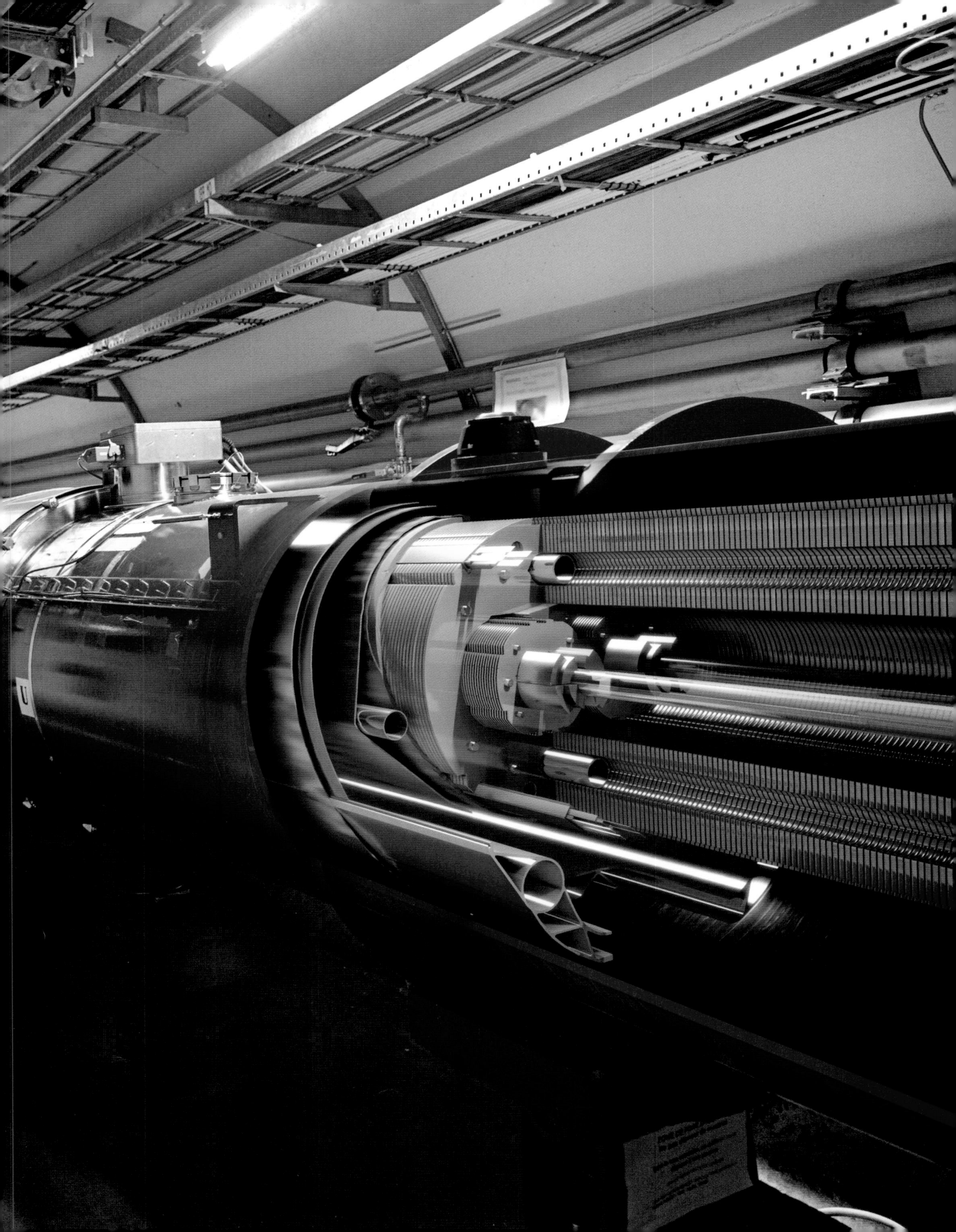

'The problem related to the high voltage supply. We get mains voltage from the grid, and there was an interruption in the power supply, just like you might have a power cut at home,' said Christine Sutton, a CERN spokesperson. 'The person who went to investigate discovered bread and a bird eating the bread.' The Large Hadron Collider had powered down, the culprit – feathers from a pigeon that had been having its meal of leftover baguette – was found at the great machine's compensating capacitor, one of several points where electricity streams through the collider 100m above the ground.

Compared to the teething troubles that the collider had previously experienced, this was nothing more than a fly in the ointment; after all, engineers had already dealt with far worse. Just two sectors of the Large Hadron Collider had heated to a few degrees above its space-cold temperature. 'There was no damage and no delay,' said James Gillies, head of communications at the particle physics laboratory. 'Had we been running [properly] we'd have lost a day or two's worth of beam time, which is nothing unusual when operating a frontier research machine like the Large Hadron Collider. Power cuts are, of course, something that the machine has been designed to cope with.'

It would be 23 November 2009 when the Large Hadron Collider particle accelerator would begin its very first run and see its very first collision between protons. Its two beams, comprising bunches of particles fired off in each direction, whizzed through the experiment with the aim of meeting at two locations within the circuit. The magnets squeezed the jets at Points 1 and 5, home to the ATLAS and CMS detectors, before tantalising ALICE and LHCb at Points 2 and 8.

It was ATLAS that got wind of the collision first. Then CMS. Then ALICE, followed by LHCb. Watching the great detectors sift through the debris of particle collisions, teams and representatives watched overjoyed in the control room. 'The past year was quite difficult,' said Fabiola Gianotti, the spokesperson for the ATLAS experiment. 'It was a big emotion.' Meanwhile, Andrei Golutvin of the LHCb collaboration proudly watched a movie of particles splashing through his team's detector to an awe-inspired crowd. 'The beam conditions are excellent,' he recalled himself saying at that moment.

The Large Hadron Collider was proving to

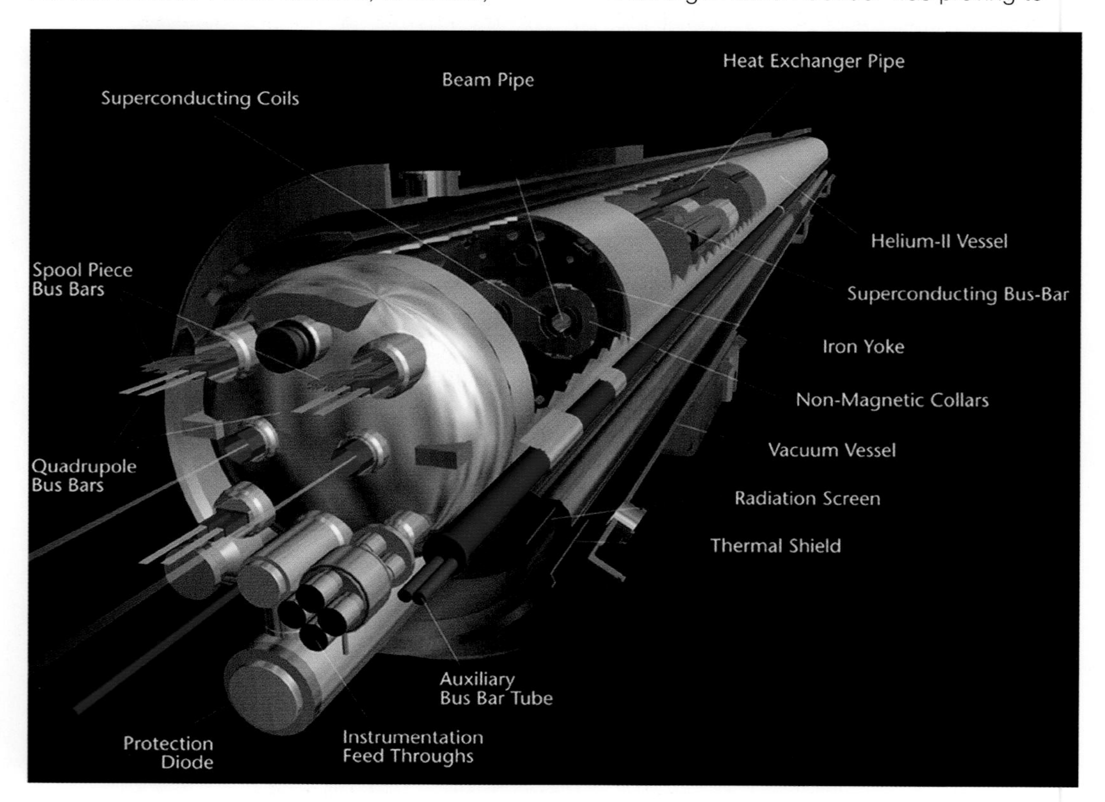

RIGHT A schematic of one of the 15m-long dipole magnets in the tube. *(CERN/AC Team)*

be fit and well, but at the time CERN weren't keen on calling their experiment a 'particle accelerator' just yet. It needed to be cranked up to its fully fledged energetic state. So far it had 'only' been smashing and crashing particles around the collider's tunnel at an energy of 1.08TeV (trillion electronvolts). It would be a few days later, on 30 November, that CERN would ramp it up to 1.18TeV. Nearing Christmastime, on 16 December, the machine was making protons collide at a much-improved energy of 2.36TeV, and the Large Hadron Collider was crowned the world's most powerful particle accelerator. With each of its four major experiments, ALICE, ATLAS, CMS and LHCb, recording over one million crashes between particles, CERN lulled the collider to sleep for the winter. Its first full period of operation was over. 'All systems were go for the Large Hadron Collider,' says Director General Rolf Heuer. 'This first running period had served its purpose fully; testing all the [accelerator's] systems, providing calibration data for the experiments and showing what needed to be done to prepare the machine for a sustained period of running at higher energy. We couldn't have asked for a better way to bring 2009 to a close.'

But Heuer wanted to go much further yet with the LHC's energy levels, to reach 3.5TeV – three times more powerful than any accelerator ever built.

The proton's journey begins

We met the Large Hadron Collider's experiments, magnets and other such components back in Chapter 2, but how do all these components fit together to achieve the particle collisions' paths of debris that played across the monitors in the control room, intensely watched by the unblinking eyes of scientists and engineers?

For part of the journey we're going to ask you to pretend you're one of many protons, a member of a bunch, inside the collider, being accelerated to speeds close to that of light. Firstly, though, you begin your journey at CERN's accelerator complex, the place where a succession of machines give you a much-needed boost.

As a proton, your life begins in a bottle of hydrogen gas. Atoms – each made up of an electron and proton – are passed through an electric field, which makes short work of the pairing; it tears the electrons away, leaving behind a gathering of positively charged protons. These are pulsed from the bottle for up to 100 microseconds per pulse; an exceedingly fast rate that ensures there are enough particles to go around. This gas-filled bottle is plugged into the end of linear accelerator 2, or LINAC 2 for short, which you'll meet next on your journey.

LINAC 2 made its debut at CERN back in 1978, since when it has worked tirelessly as the laboratory's major source of proton beams. Its job isn't too dissimilar to its predecessor, LINAC 1, which was mapped out and built during the 1950s: it deals with 50MeV protons, handing them over to the Proton Synchrotron Booster. During your journey in the accelerator you'll also be introduced to the radio frequency cavities, the metallic chambers where an electromagnetic field is made which charges up several cylindrical conductors.

Now forget you're a proton for a moment: stand outside these cavities and they'll remind you of a beaded necklace – the beads are the cavities and the string that connects them is the tube through which you'll be travelling in a moment. Back inside your tunnel, it's like being in space – you're surrounded by a vacuum. While you wait to begin your journey you need an electromagnetic field behind you, which is where you get a helping hand from a radio frequency power generator.

Every component on the Large Hadron Collider, right down to the last nut and bolt, has a specific purpose – to get the job done, by working together to solve the myriad mysteries of the universe. The RF cavities are no exception; they have a specific size and shape that allows waves of a freshly generated electromagnetic field to build up inside them.

You're now ready to begin your journey through this part of the LINAC 2. Being a charged particle you feel the force behind you, and you're pushed through the tube. During your travels you'll realise that your timing needs to be perfect, because you need to be sorted into a group with your fellow protons.

As was mentioned earlier, particles make their way around the Large Hadron Collider in gangs, or packets, of some billions of protons. In order to 'make the grade' and join the group, the field in any given radio frequency cavity likes to change direction – or oscillate – quite regularly, at a frequency of about 400MHz. That's 23,900 times higher than the tone a television cathode ray tube makes whilst it's running. Protons that are punctual and arrive at the next step in a timely fashion have an accelerating voltage that's equal to zero, but not all particles get it right; if you have an energy that means you arrive too early or too late for the party, you'll either be sped up or slowed down, whichever is required to ensure you're an 'ideal' particle that can be sorted into your bunch.

The cylindrical conductors in which the RF cavities work their magic allow you to pass through them. They have alternating positive or negative charges. Of course, having a positive charge yourself you're going to be repelled by a positive conductor and attracted by an oppositely charged one, so you're pushed and pulled along, which causes you to accelerate, while a quadrupole magnet – that's a group of four magnets laid out in a ring, north next to south which is next to north, where another south pole completes the arrangement – generates a field that forces you to stay in line with your fellow protons in the shape of an exceedingly tight beam. Because you're moving faster you've gained about 5% in mass, and are now ready to enter the Proton Synchrotron Booster.

As its name suggests, the Proton Synchrotron Booster is the section in the Large Hadron Collider that gives protons an extra kick of energy, which they need to gain admission to the next step in the sequence. The booster is the first circular proton accelerator in the chain, and also the smallest; but its collection of four synchrotron rings measures 50m across and is stacked one on top of the other. Inside each of these you'll find a set of quadruples and dipoles, the latter having north and south poles on opposite sides; the simplest example is a standard bar magnet. In the LHC's arrangement there are 32 dipoles and 48 quadrupoles, each making up the vertical stack with a common yoke that allows one main power supply to blast every magnetic member of the series with a current.

Packed into the booster is an elaborate collection of correction loops that have the job of focusing the fields in each of the four rings. This is where your life as a proton continues; you've left linear accelerator LINAC 2 behind and you're now part of a particle packet that enters the booster vertically, or straight up, before being multiturn-injected into each of the rings. It's here where you race around each one, thrust along and steered by the inbuilt magnets at 91.6% the speed of light. You're also jammed closer to other protons in your packet. Such is its influence, the Proton Synchrotron Booster is able to smash a 50MeV beam with such intensity and speed that it's thrown out with an energy of 1.4GeV. When the job's done the booster throws you through one of two back doors, either to the Proton Synchrotron or to the On-Line Isotope Mass Separator, also known as ISOLDE.

ISOLDE is like a small alchemy factory; it turns one element into another. It plays with quite an unusual source of beams made up of very radioactive isotopes – in other words, the nuclei of atoms that don't have enough neutrons to be in any way stable. Getting these atomic fragments is a matter of using the help of the high-intensity proton beam that races from the Proton Synchrotron Booster before hitting a specially-developed target, dependent on the type of isotope that physicists wish to create. An array of devices lie in wait, ready to strip electrons from these atom pieces before extracting and sorting them according to their mass. A low-energy beam is then delivered to a selection of experimental stations or to the recently implanted linear accelerator, the High Intensity and Energy ISOLDE (HIE-ISOLDE), which energises the radioactive beams up to about 7.5MeV per nucleon, with plans to upgrade this accelerator further, which will see the machine accelerate nuclei up to 10MeV per nucleon. These accelerated beams make their way to the aptly named Miniball station, an array of detectors composed of highly pure germanium.

Being a dab hand at shape-shifting members of the periodic table, ISOLDE's résumé is actually quite impressive. To date it has produced about 1,000 different isotopes for particle physicists to study up close. It's also made 600 radioactive variants of 60 elements, from colourless helium gas through to silvery-white radium, which disintegrate in a matter of mere milliseconds.

Now, look around you. You're in the Proton Synchrotron, and you're going to be accelerated to much higher speeds – up to 99.9% the speed of light, in fact. This synchrotron, which accelerated its very first particles back in November 1959, is still going strong, whirling protons around in a system 628m in circumference packed out with electromagnets to a whopping energy of 25GeV.

Inside the synchrotron you'll be pushed along by cavity resonators. It's here where electromagnetic waves at a standstill are made to move in 'tune' with a resonator. Think of playing a musical instrument like a recorder: when you blow into it, the wood begins to vibrate and a turbulence with a range of vibrational frequencies is created. Eventually the frequency at which the instrument is vibrating matches up with that of the column of air and the two make a sound together. The cavity resonators work in the same way, putting the electromagnetic waves in a resonance and thereby accelerating the protons forward.

The electromagnetic waves accelerating you are microwaves, and inside the Proton Synchrotron you make the most of them, utilising the cavities tuned to the peaking waves that are characteristic of the electromagnetic spectrum. If the resonant frequency of one of these cavities matches up with the frequency that's being stimulated, then you get peaking, which in turn creates a field strength greater than that possessed by the microwave radiation we met earlier. This is where you get your kick-start around the accelerator; but there's a difference between the two, and that's what causes you to race around the synchrotron – very much like a surfer riding a huge wave on a surfboard.

The Proton Synchrotron is affectionately known as CERN's beating heart. Not only does it accelerate particles, but it's also known as the multi-tasker of the accelerator complex. It's also a juggler of particles, sorting through not just protons but also, in its past, alpha particles – that's the nuclei or centre of helium atoms without the electrons whizzing in orbit around them – as well as oxygen, sulphur nuclei, electrons on their own, antiprotons and positrons. Although not all at the same time!

In terms of 'your' particle, the synchrotron gets to work on six proton packets that have been passed on from a two-step

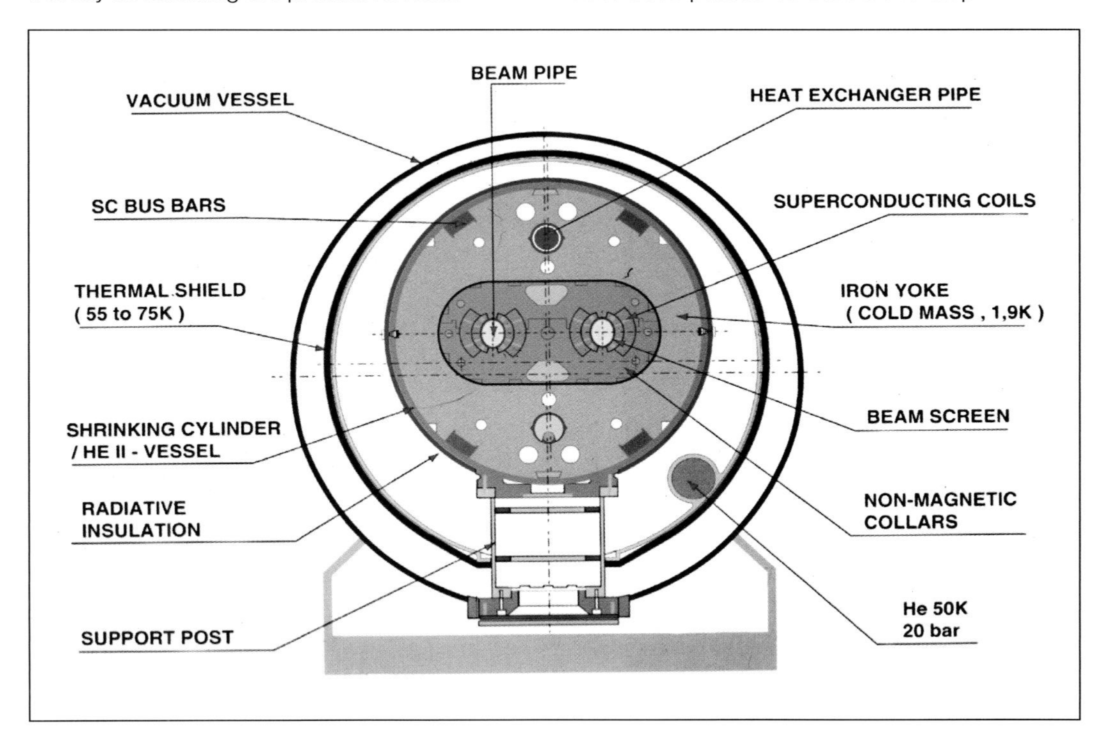

LEFT A cross-section of one of the dipole magnets that accelerate protons around the LHC. *(CERN)*

booster, slicing them into 72 packages, just 4 nanoseconds long and separated into 25-nanosecond intervals. How it does this is a matter of ramping up the frequency – the Proton Synchrotron Booster snatches the six packets from the Proton Synchrotron at 3.06MHz, switching off the power they supply and simultaneously hitting them up several notches to 9.10MHz. At the moment 16 packages have been made; it's not until the protons are accelerated at 25GeV, kicking up the frequency to 10MHz and then 40MHz, that a little over 70 particle packets are produced. While the Proton Synchrotron has a need for speed as far as particles are concerned, it does like to leave a 320-nanosecond gap between each couple of packets. This gives its kicker magnet enough time to take the beam out of its circular orbit in the accelerator before spitting it out into the next stage in your journey – getting splashed around in the superbly named Super Proton Synchrotron.

An accelerator within an accelerator

At almost 7km in circumference, the Super Proton Synchrotron is the second-biggest machine within the complex where CERN's accelerators are housed. Built to race particles to energies of 400GeV, it's the beam-weaver for an array of machines besides the Large Hadron Collider, for it has some experiments of its own, including the Heavy Ion and Neutrino Experiments, known as NA61/SHINE and NA62, as well as the Common Muon and Proton Apparatus for Structure and Spectroscopy (COMPASS).

NA61/SHINE's interests lie in getting up close and personal with hadrons, the particles that take part in strong interactions. These members of the subatomic family keep quarks locked together and stop the constituents of atomic nuclei from flying apart in all directions. The experiment is interested in picking apart how hadrons are made, particularly when particle beams – made out of protons, pions, beryllium nuclei or argon and xenon nuclei – crash together. The Super Proton Synchrotron is the supplier of 400GeV protons, which smash into a plate coated in beryllium some 500m away from the NA61 experiment. There's a beam line, known as H2, which picks out hadrons with a certain momentum or charge that gets transported to NA61.

To make sense of the hadrons, the NA61 employs four Time Projection Chambers, a type of detector that uses a melee of electric fields, magnetic fields and a volume of gas to create a three-dimensional picture that reveals where the particles are going and what happens when they smash into each other. These chambers get a helping hand too, in the form of Time of Flight scintillator arrays that assist in the identification of particles. These are able to figure out which particles are light and which are the heavier members of the elementary family, by working out how long it takes them to race between two scintillators. Think of it as timing a relay race, where the first runner activates a clock while the next stops the timer when the particle smashes into it. Having an idea of the duration of its flight and the speed at which the particles are racing from one scintillator to the next gives NA61 information on the particles it's dealing with. Meanwhile, inside the same experiment, you'll find the forward calorimeter, whose job is to investigate the energy at which subatomic fragments fly off in all directions immediately after beams collide. The set-up is completed with another, albeit small, Time Projection Chamber and beam definition detectors.

It's a robust experiment, keen to get the job done with its arrangement of detectors. Using these, NA61 is able to work out where hadrons are located, how heavy they are and whether any particles smashed into the beryllium plate at all. The team behind the experiment are then able to choose a specific hadron type – for example, resulting from the collision of proton and pions and other such beam flavours – and get information from NA61 on how many hadrons were made and the momentum they have, as well as their mass and charges. The experiment is then able to delve deep into getting a clearer picture of the line that separates quark-gluon plasma – the soup that's thought to have existed just after the birth of the universe – and gas dominated by hadrons.

The Super Proton Synchrotron's other assistant is the kaon-obsessed NA62. It hasn't been at work for a great deal of time, since it

ABOVE Repairing one of the dipole magnets in the LHC tunnel. *(CERN/Maximilien Brice)*

was only approved by CERN back in February 2007 and has only recently begun taking data on the rare decays of charged kaons, which ultimately allows the machine to make precise tests of the Standard Model; more specifically, particle interactions that happen at exceedingly small distances. You might have heard of the kaon being referred to as a K meson. That's because it's made up of four of these hadronic subatomic particles. Break them down even further and you'll find a quark and its opposite, the antiquark. NA62 takes advantage of this by getting some idea of how top quarks fall apart.

Particle physicists know that kaons disintegrate into a charged pion as well as a neutrino and antineutrino pairing, a piece of information that provides some headway into getting to grips with mysteries on a more subatomic level. To do this, though, NA62 needs plenty of kaons to play with. To get them it works on the same kind of principle as fellow detector NA61. Being the beam supplier, the Super Proton Synchrotron fires high-energy protons into a target entirely made up of beryllium. Protons meet beryllium atoms at high velocity and almost one billion particles per second are made, though only about 6% of these are the candidates NA62 is after: kaons.

What the kaon factory does next is a story of high-resolution timing, detectors, trackers and other such hardware. The experiment also makes use of a huge tank filled with a vacuum. First, however, the speeding particles need to be checked over or measured by a detector that has pixels made up of silicon. This is the GigaTracker, which picks up important information about the 75GeV kaon beam fired its way, chiefly to work out the time, direction and momentum of the beam tracks. This detector doesn't comprise just one station – it has three, mounted close to four magnets that work continuously to ensure that the beam is sufficiently deflected. The GigaTracker itself sits inside a vacuum, operated at a temperature of about -20°C (-4°F).

Being aware of the identity of each and every particle in the beam is also very important. For this the NA62 employs CEDAR (Combined e-Science Data Analysis Resource), a counter that enjoys watching light shows in order to make sense of its target. CEDAR comprises a pressure vessel that's filled with hydrogen gas, maintained at a precisely controlled pressure of 3.85 bar; that's almost four times the pressure you're experiencing from our planet's atmosphere as you sit and read this book. Right at the end of the vessel, illumination – known as Cherenkov radiation – is reflected by a curved mirror on to a ring-shaped diaphragm that measures about

100mm at its beginning. Cherenkov radiation is rather like a sonic boom, only with light rather than sound. It's a characteristic flash of blue light that's brought about when a charged particle races through a medium at a speed quicker than the usual speed of light through the same material. Since the medium can't react to the speeding particle fast enough, a disturbance – or shockwave – is left behind, trailing after the particle that made it.

Inside the tank is where the particle decay magic happens as detectors look on. Their job is to seek out the particles that are the result of kaon decay. In particular, the Ring Imaging Cherenkov (RICH) detector makes the most of the 'light boom' to help it identify the electrically charged subatomic particles that become the aftermath of a kaon's decay. A sizeable system of photon and muon detectors spits out any decays NA62 doesn't want, never to be looked at again.

NA62's Straw Tracker is, as its name suggests, made up of thin straws that measure the direction and momentum of particles that shoot from the decay region. There are over 7,100 of these 'straws' to be exact, and they work as tiny drift chambers through which particles pass and which tear off electrons from gas molecules inside, which in turn let out a signal. These straws are also quite sturdy and leakproof, capable of withstanding the enormous pressures of the gas inside the 100m by 2.5m vacuum tank, and are created by winding a pair of conductive tapes in a spiral. These pack the insides of four chambers, which are cut into by a dipole magnet so that the detector is able to catch the particles in four different directions. That way it can provide four coordinates.

Investigating how quarks and gluons get along to provide the zoo of particles we're able to observe – from the simple proton all the way to the most exotic members of the family – the 60m-long Common Muon and Proton Apparatus for Structure and Spectroscopy (COMPASS), which also goes by the name of NA58, is the last of the synchrotron's experiments. One of its major aims is to find out more about the property of spin, which we met in Chapter 1, and how it comes about in the proton, and its neutral-nucleus companion the neutron. In particular, it asks how much spin is contributed by the gluons that weld quarks together through the strong force. To find out, COMPASS has three major components: a beam telescope, a two-staged spectrometer and a target area.

Muons – the elementary particles that aren't too dissimilar to the negatively charged electron, only with a bigger mass – speed from the Super Proton Synchrotron while a primary proton beam, packed with 400GeV of energy, is guided towards a beryllium production target that's approximately 1.1km away on what's known as a transfer line. As soon as the beam strikes it, the target produces hadrons made up of a selection of antiprotons, pions and kaons.

When it comes to gluons, COMPASS is also interested in the range of particles that they and quarks can make up. Instead of muons, though, the experiment needs to use a beam of pions to investigate.

We've detoured a little from the Large Hadron Collider. However, the Super Proton Synchrotron, while it has its own particle physics experiments working away, also works for the particle accelerator. It's been the workhorse of CERN's particle physics programme for quite a few decades, since it was switched on for the very first time in 1976. Its circular tunnel almost seems to cut into that of the Large Hadron Collider, since it too straddles the border between France and Switzerland. You'll remember the Large Electron-Positron; the Super Proton Synchrotron was the injector for this particle smasher.

Now – and ever since 2006 – the accelerator makes use of over 1,300 room-temperature electromagnets, of which 744 are dipoles, that allow it to bend the particle beams smoothly and quickly around its ring. Having its own experiments the synchrotron has dealt with an impressive selection of particles, including electron-less sulphur and oxygen, as well as electrons and their antiparticle, the positron. When it comes to the Large Hadron Collider, though, the proton and the antiproton are its main interest.

Let's pick up from where we left off in our journey as a proton to the Large Hadron Collider. As you're thrust around the synchrotron ring at breakneck speeds you'll feel the strength of the magnets that squeeze and push you on your way. You're getting ready to be injected into the particle accelerator's main set-up.

Before you do, though, you'll get a choice of pipe to race through. These are the pathways that determine which direction you'll move around the accelerator – that is, either in a clockwise or an anticlockwise motion. Whichever tube you choose, you'll feel a vacuum around you. That's to ensure that you don't knock into any gas molecules, and it's like being in a world that's as empty as outer space. It's not the only vacuum you'll meet during your travels – it's needed to insulate the magnets, cryogenically cooled to -271°C (-456°F – that's colder than space, which is a frigid -270°C/-454°F), and is put in place to ensure that not too much heat from the surrounding room temperature gets into the tubing. It's the same story for the distribution line, which carries liquid helium. In order for the Large Hadron Collider to operate as it should, keeping cool is very important. In fact, without the cryogenics – which makes use of 40,000 leak-tight pipe seals, 40MW of electricity and 120 tonnes of helium – the superconductivity that's needed for the wires alone, a mishmash of niobium and titanium metals, wouldn't be possible.

The Large Hadron Collider has a cooling system of as many as five cryogenic islands. Their job is to ensure that coolness reaches kilometres of magnets. There are three cooling stages that ensure the accelerator is as cold as can be; it also takes several weeks to reach optimum temperature. Firstly, 10,000 tonnes of liquid nitrogen needs to be cooled to -193°C (-315°F). This is then injected into a cooler collection of magnets and fanned by turbines that cool it even more, to -269°C (-452°F). Finally, third and fourth stages equipped with refrigeration units bring the temperature down to -271°C (-456°F), and some 36,000 tonnes of magnets have been cooled. Why CERN chose helium as a coolant is all down to the properties it owns: it's able to keep everything cool over long distances, is a brilliant heat conductor, and is an all-rounder in the refrigerant department.

All in all, the Large Hadron Collider has one of the largest vacuum systems in the world. Not only is it cool, it's kept at a pressure that's 1.014 billion times less than the air pressure you're feeling right now, throughout 50km of piping of 15,000m^3. For one of the greatest vacuum systems in the world, you'll need seals and joints that the Large Hadron Collider has in spades – 250,000 and 18,000, respectively to be exact. Travel down the remaining 54km of pipes and you'll feel a similar vacuum to what's found on the lunar surface, that's ten trillion times less than the Earth's atmospheric pressure. This is all topped off with a thin titanium-zirconium-vanadium coating that sucks up any residual molecules that may appear, including methane and other more inert gases that you'll recognise as the noble gases in the Periodic Table. These are then forced out by a sequence of 780 ion pumps.

After speeding through the tubes, you've now entered the particle accelerator's 27km ring. As a proton, a strong magnetic field will serve as your guide, directing you in a circular path by a sequence of superconducting electromagnets. As of 2018 the Large Hadron Collider is spitting out beams at 6.5TeV apiece, which means that if you don't crash into a proton coming in the opposite direction it'll take you one second to complete over 11,200 laps around the tunnel; and, as you speed along, you'll come across more than 50

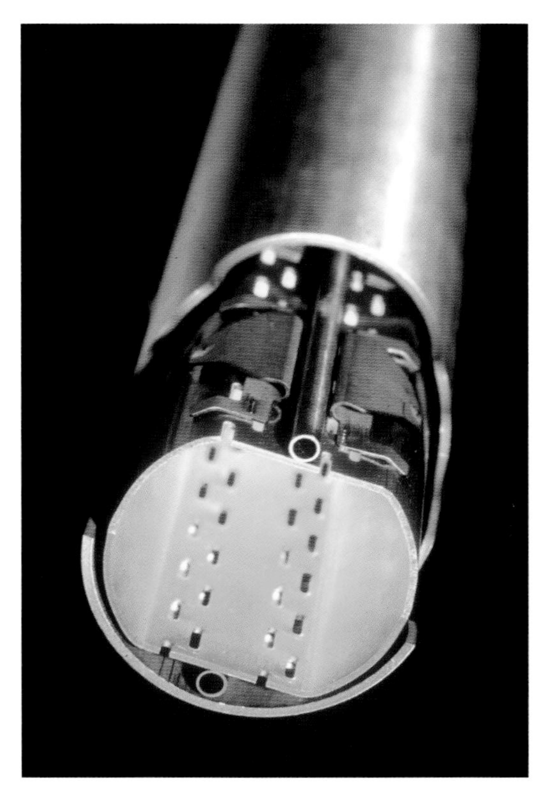

LEFT A cross-section of the beam pipe, through which the protons travel and collide. The screen lining the inside of the pipe has slits that allow gas molecules to be pumped out. *(CERN/Patrice Loïez/ Peter Rakosy/ Laurent Guiraud)*

kinds of magnet to ensure that you don't slow down – not even for a millisecond. Which is why, when considering electromagnets for the particle accelerator, CERN had to think power – more than 100,000 times that of the Earth's magnetic field. Only then would they be able to achieve a consistent 8.3 teslas. These magnets need to remain magnetised, and to ensure this a current of 11,000A creates the field, whilst a superconducting coil allows those high currents to flow.

There are thousands of what are known as lattice magnets, which assist with bending and tightening your trajectory around the Large Hadron Collider. Without them it's not possible to keep the beams stable and ensure that they're aligned just right for a collision. In the LHC arrangement you'll find over 1,200 dipole magnets – each some 15m long and weighing in at 35 tonnes – which, combined with an intense magnetic field, ensure that, as a high-speed particle, you're able to take care of those very tight turns along your path whenever you need to. In fact, such is the power of the arrangement that if you were to use normal magnets like the ones you find in your fridge, but hundreds of times stronger, the Large Hadron Collider would need to be just over four times longer to match the same energy.

Of course, it's the quadrupole magnets – of which there are almost 400, between 5m and 7m in length – that keep the beam nice and neat. You'll go through these as you move through the beam pipe, where they're squeezing the beam both vertically and horizontally for good measure. In addition, the tighter or more organised your proton packet is the more likely it is that you and your fellow protons will strike it lucky when it comes to crashing into another set of particles. If their magnetic field isn't quite perfect in places then the Large Hadron Collider's system of magnets makes short work of it; the dipoles have sextupoles, octupoles and decapoles ready and waiting to get that particle beam shipshape. Just for reference, the Large Hadron Collider's technology is so precise that if it were to fire two needles towards each other at a distance of 10km they'd meet – head-on – at the halfway point.

BELOW Schematics of 'A Large Ion Collider Experiment' (ALICE). *(CERN)*

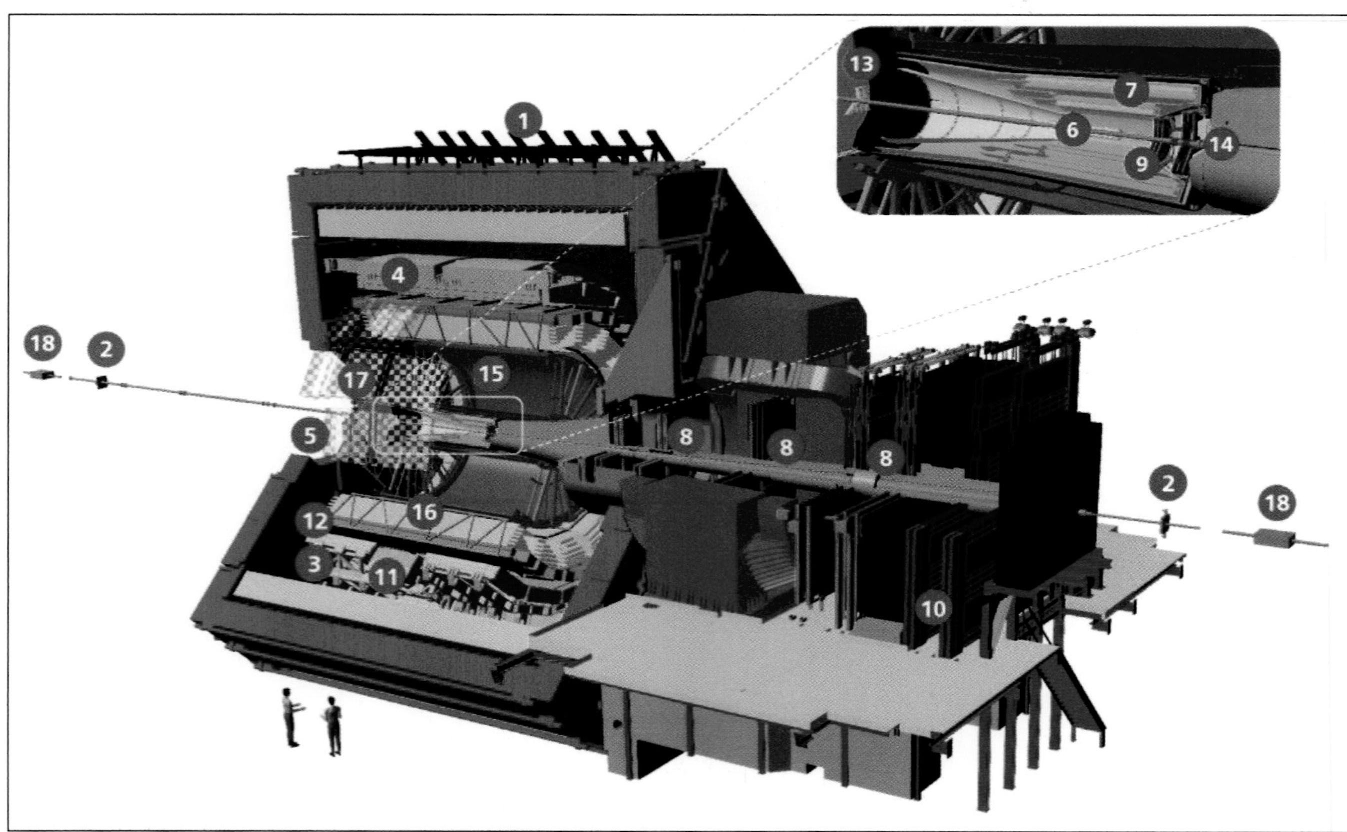

Meeting the experiments

Of course, during your journey along the ring of the Large Hadron Collider you're going to meet all the main experiments – ATLAS, CMS, ALICE and LHCb. And, as you've probably already guessed, it's magnetic magic that ensures you're on the straight and narrow. This is generated by the insertion magnets, three quadrupoles that make up a triplet that squeezes the beams even closer together – some 12.5 times narrower (from 0.2mm to 16µm across) – before they make their way through the detector. It's around this point that some familiar protons, the ones made with you in LINAC 2, will be found coming at you from the opposite direction, ready to strike. You'll pass eight inner triplet magnets on your way around the Large Hadron Collider if you're not obliterated in one of the very first collisions. There are two assigned to each of the four detectors and they also have the important job of keeping the beams clean; if stray particles made their way into the particle accelerator's most sensitive equipment then the data would be worthless.

1. **ACORDE** | ALICE Cosmic Rays Detector
2. **AD** | ALICE Diffractive Detector
3. **DCal** | Di-jet Calorimeter
4. **EMCal** | Electromagnetic Calorimeter
5. **HMPID** | High Momentum Particle Identification Detector
6. **ITS-IB** | Inner Tracking System - Inner Barrel
7. **ITS-OB** | Inner Tracking System - Outer Barrel
8. **MCH** | Muon Tracking Chambers
9. **MFT** | Muon Forward Tracker
10. **MID** | Muon Identifier
11. **PHOS / CPV** | Photon Spectrometer
12. **TOF** | Time Of Flight
13. **T0+A** | Tzero + A
14. **T0+C** | Tzero + C
15. **TPC** | Time Projection Chamber
16. **TRD** | Transition Radiation Detector
17. **V0+** | Vzero + Detector
18. **ZDC** | Zero Degree Calorimeter

Finally, the moment has arrived. You're about to take part in a collision; but your journey isn't over just yet. After crashing into your partner at a speed close to that of light, you start to fly apart. It's these fragments that the detectors of the Large Hadron Collider are interested in. They tell the detectors everything about you; everything you were and everything you are now after being smashed to smithereens, like your mass, your charge, your speed and your structure.

Ready to pick up your fragmented pieces, gigantic magnets are ready to assist in the beam-on-beam measurements. Information such as how the charged particles' response to the magnetic field provides clues to your particle's identity, while your momentum can be determined from the angle at which you were deflected. Meanwhile, dipole magnets thrust the particle beams apart again, whilst other magnets clip away any particles that seem to have spread out after the impact. More often than not the Large Hadron Collider needs to dispose of particles, and the ones it selects for elimination are guided down to a dumping ground for beams, where a dilution magnet awaits. Here the beam's intensity is stripped down by 100,000 times before it crashes into a block of concrete. The particles have been killed off.

ALICE's aims

If you take a clockwise direction around the Large Hadron Collider, the first detector you'll meet is ALICE. This impressive experiment – which, as you've already discovered from Chapter 2, is encased in a red, octagonal magnet inherited from the Large Electron-Positron collider – identifies particles. It likes to work out whether it's dealing with photons, pions, electrons or even protons in the aftermath of a beam collision. It also likes to work out their charge and mass in a hot volume before such heated conditions have a chance to cool down, catching them in its detectors before they disappear for good. ALICE has an impressive array of 18 detectors. The first comprises a tracking system.

Around the interaction point, detectors in the shape of barrels have a front-row seat to

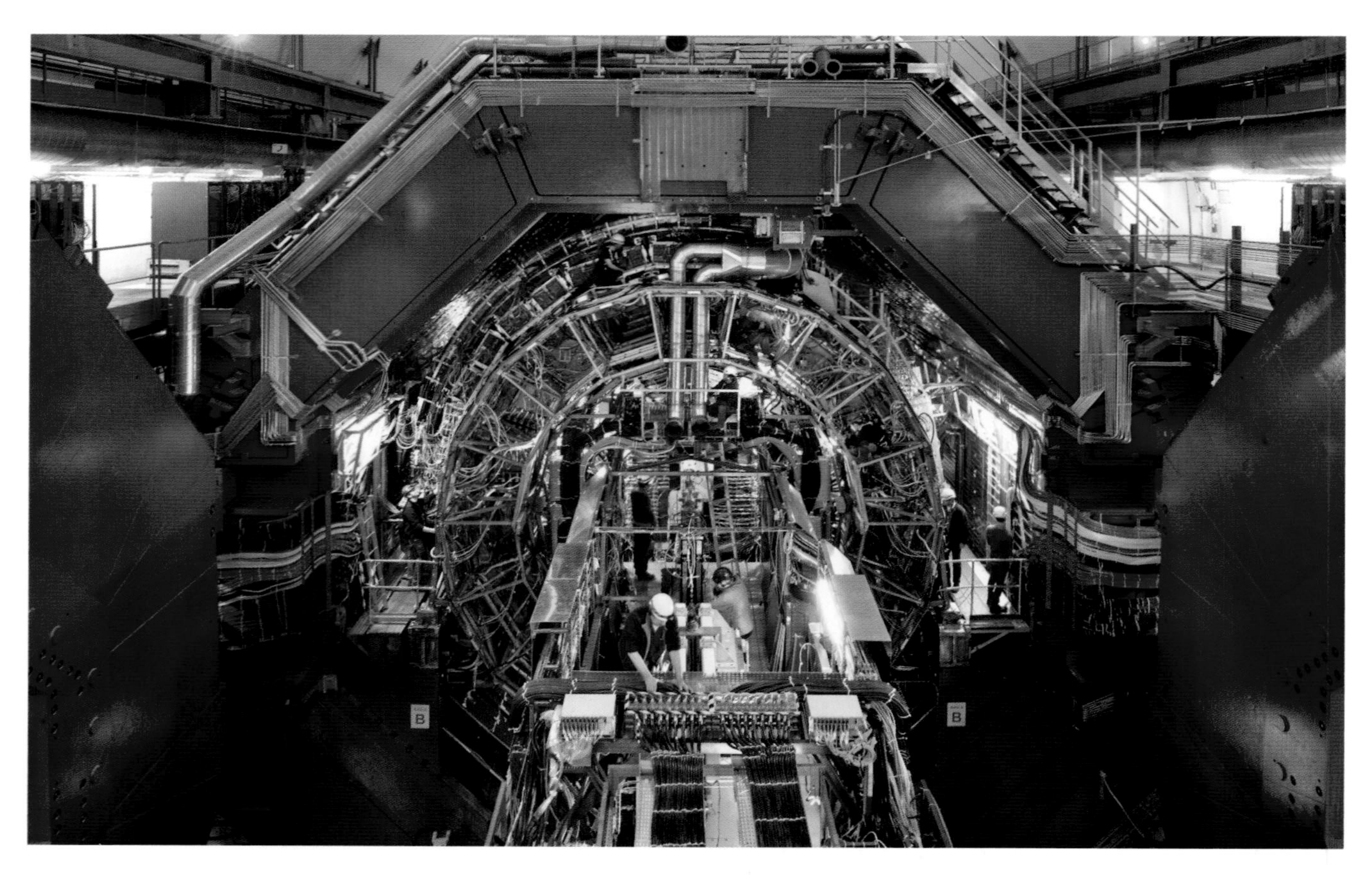

ABOVE An engineer works at the Large Hadron Collider's ALICE experiment. *(CERN/A. Saba)*

RIGHT Members of the ALICE collaboration. *(CERN/Maximilien Brice)*

the particle-smashing action. They're lying in wait for any runaways that manage to escape the hot, dense 'plasma' that the Large Hadron Collider has cooked up. They're known as the Inner Tracking System (ITS), the Time Projection Chamber (TPC) and the Transition Radiation Detector (TRD). All of the detectors are able to give precise information on each particle's whereabouts and whether they're positively or negatively charged, nestled in a magnetic field of 0.5 teslas that give them their precision, to a whisker of about a tenth of a millimetre.

The Inner Tracking System is made up of six cylindrical layers of silicon detectors. Each and every layer encompasses the collision point, ready to find out more about the particles that emerge the other side. To this tracking system in particular, working out where the heavy members of the quark family – notably charm and beauty – are is like child's play; two layers of Silicon Pixel Detector, followed by a couple of layers of Silicon Drift Detector, and finished off with two layers of Silicon Strip Detector, provide it with all the tools it needs to reconstruct what

LEFT ALICE's Time Projection Chamber detector readout chamber. *(CERN/Hans Rudolf Schmidt)*

BELOW A technician talks a stroll on a walkway beside the giant ALICE experiment. *(CERN/Anna Pantelia)*

happens when they fall apart in their decay. The Inner Tracking System is the largest of ALICE's detectors, taking over an area of 5m^2 at the very heart of the Large Hadron Collider's experiment.

The device that's the head honcho of particle tracking is the Time Projection Chamber, which uses a mix of magnetic and electric fields to map out their path. It also makes use of a gas chamber, the contents of which make the particles' path obvious as they race through the gaseous atoms, stripping them of their electrons, which then drift towards plates at the end of the detector. How strongly a particle rips electrons out of the gas gives a good indication of the culprit responsible for the ionisation, revealing an important snippet of information to the ALICE experiment. When the charged particles crash, they ooze energy – something that the experiment's detectors never miss.

A counter helps ensure that ALICE doesn't miss a trick. This is achieved by means of an avalanche of ions made around a series of wires inside the 'readout' chamber. The wires are anodes, which suck up the deluge of positive ions, amplifying the signal, which in turn induces a positive signal on a pad. In

OPPOSITE TOP The Semi-Conductor Tracker (SCT) ready to be inserted into ATLAS' Transition Radiation Tracker (TRT). *(CERN/Maximillien Brice)*

OPPOSITE BOTTOM The Semi-Conductor Tracker entered into ATLAS' Transition Radiation Tracker (TRT). *(CERN/Serge Bellegarde)*

ABOVE An exterior view of the Semi-Conductor Tracker as it is being lined up to enter the Transition Radiation Tracker during ATLAS' construction. *(CERN/Serge Bellegarde)*

BELOW Detail of cables feeding into the ALICE instrument. *(CERN/Anna Pantelia)*

order to get the readout, there are cathodes at work inside a particle-detecting multi-wire proportional chamber that's able to track the trails of ionisation in the gaseous chamber and work out the position of a speeding particle. These counters can be found at the Time Projection Chamber's end plates. And there you have it; the radial distance, azimuth (or angle) and the beam direction – which can be quickly calculated by the particle's drift time – can be used to make a three-dimensional map of what's happening when beams collide.

ALICE also needs to work out whether it's dealing with electrons or their positive counterparts, positrons, from other charged members of the particle family during the beam collision process. For that, the experiment gets a helping hand from the Transition Radiation Detector, which is able to keep a detector-eye out for any X-rays that stray through its layering of material. Since different charged particles are able to release differing amounts of electromagnetic radiation when they encounter the many sheets of materials of different thickness, telling the particles apart is a piece of cake for the Large Hadron Collider's detector experiment – even though the effect is tiny. Photons carrying X-ray energies can be found just behind the boundaries of material, where they're detected using the multi-wire proportional chambers filled with xenon gas and drop their energy right on top of the electron-shredding signal that was pulled from the particle track. The Transition Detector is quick, sorting through the kinds of high-momentum particles it meets and working out how much energy is linked to them with the help of 250,000 central processing units.

Speaking of particle identification, ALICE is able to distinguish between electrons, protons, kaons and pions in the aftermath of a collision by means of a strategic plan. It's easy to work out the mass of a particle from its momentum

BELOW Tracks in ALICE's Time Projection Chamber. These particle paths are a cosmic shower event from the muon absorber. *(CERN)*

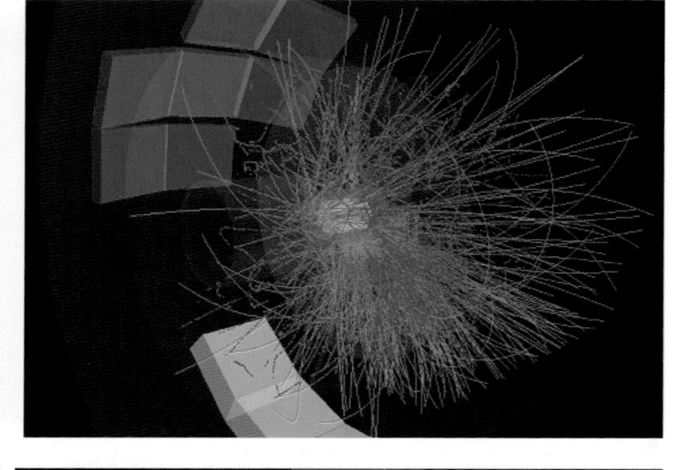

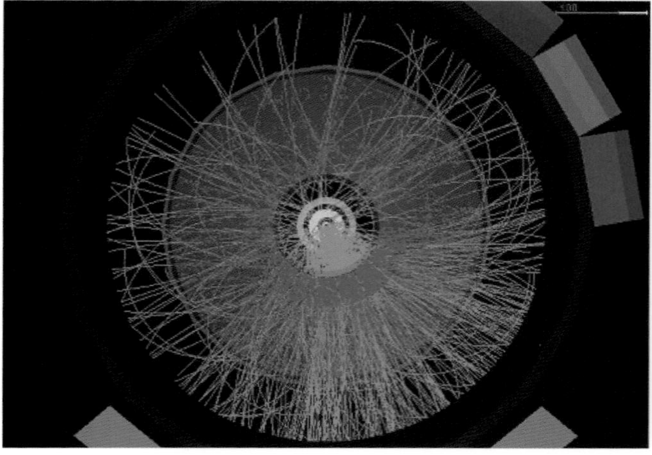

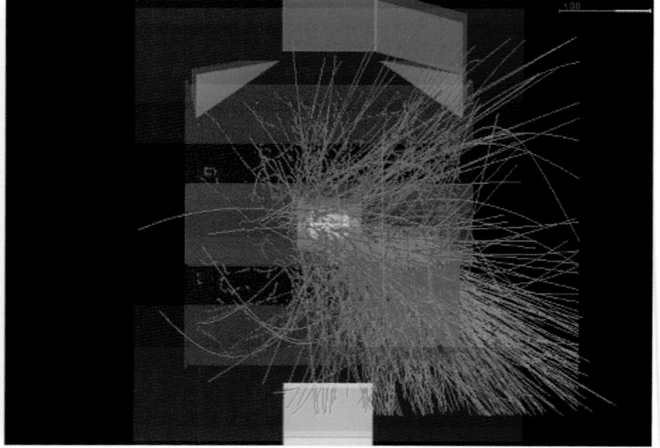

ABOVE **ALICE's Time Projection Chamber.** *(CERN/Hans Rudolf Schmidt)*

LEFT **Supermodules of the Time-of-Flight detector on ALICE.** *(CERN/A. Saba)*

BELOW LEFT **ALICE's High Momentum Particle Identification (HMPID) gas distribution system.** *(CERN/A. Saba)*

BELOW **ALICE's multi-layer cooling pipes.** *(CERN/A. Saba)*

as well as how fast it's moving, while locking down the momentum and whether the particle possesses a positive or negative charge goes hand-in-hand with how much it follows a curved path in response to the magnetic field coursing through the particle accelerator and its detectors. Meanwhile, working out the speed at which the particles race around after hatching from the collision wouldn't be possible without the help of time-of-flight and ionisation measurements, combined with sampling transition and Cherenkov radiation.

That's not always enough to fully grasp the bits of information found in the particles

LEFT ATLAS stands 25 metres tall. This shows the view of the experiment's eight toroidal magnets during construction. *(CERN/Maximilien Brice)*

created by the Large Hadron Collider. To achieve this much more specialised detectors are needed: a Time-Of-Flight instrument and a High Momentum Particle Identification Detector. As you've more than likely guessed, the former measures the time it takes each and every particle to 'fly' away from the interaction point. Meanwhile, the latter seeks out the faint light patterns that are made from super-speedy particles, whilst the Transition Radiation Detector works in unison, catching sight of special radiation that the very speedy particles emit, picking out the electrons and identifying the muons when they race through any matter they encounter. We'll come back to muons further on in this chapter.

ALICE's Time-Of-Flight detector has a large cylindrical surface area, some 141m^2, with an inner radius of 3.3m. It makes use of 160,000 plate chambers, multigap resistive and with time resolution of approximately 100 pico-seconds (that's one trillionth of a second). Look closely at the multigap resistive plate chambers and you'll see that they're made of thin sheets of glass, rather like you'll find constituting the windows of a house. These parallel plates create narrow gaps of gas where intense electric fields race, and are separated at the perfect spacing with the help of fishing lines. For an efficiency that's as close to 100% as can be, ten gaps of gas are needed.

Ring-Imaging Cherenkov, or RICH, detectors are required by ALICE in order to work out the speed of particles that have a momentum too great to be captured by the Time Projection Chamber and Inner Tracking System or from time-of-flight measurements. The High Momentum Particle Identification Detector (HMPID) is perfect for such a job and, being a member of the RICH detector family, uses the aid of Cherenkov radiation to get at the mass of a particle by measuring its speed. A photon detector a mere 10cm from where the packets of Cherenkov light spill ensures that the blue cone of light that's so characteristic of the effect expands, pulled into a ring-shaped structure.

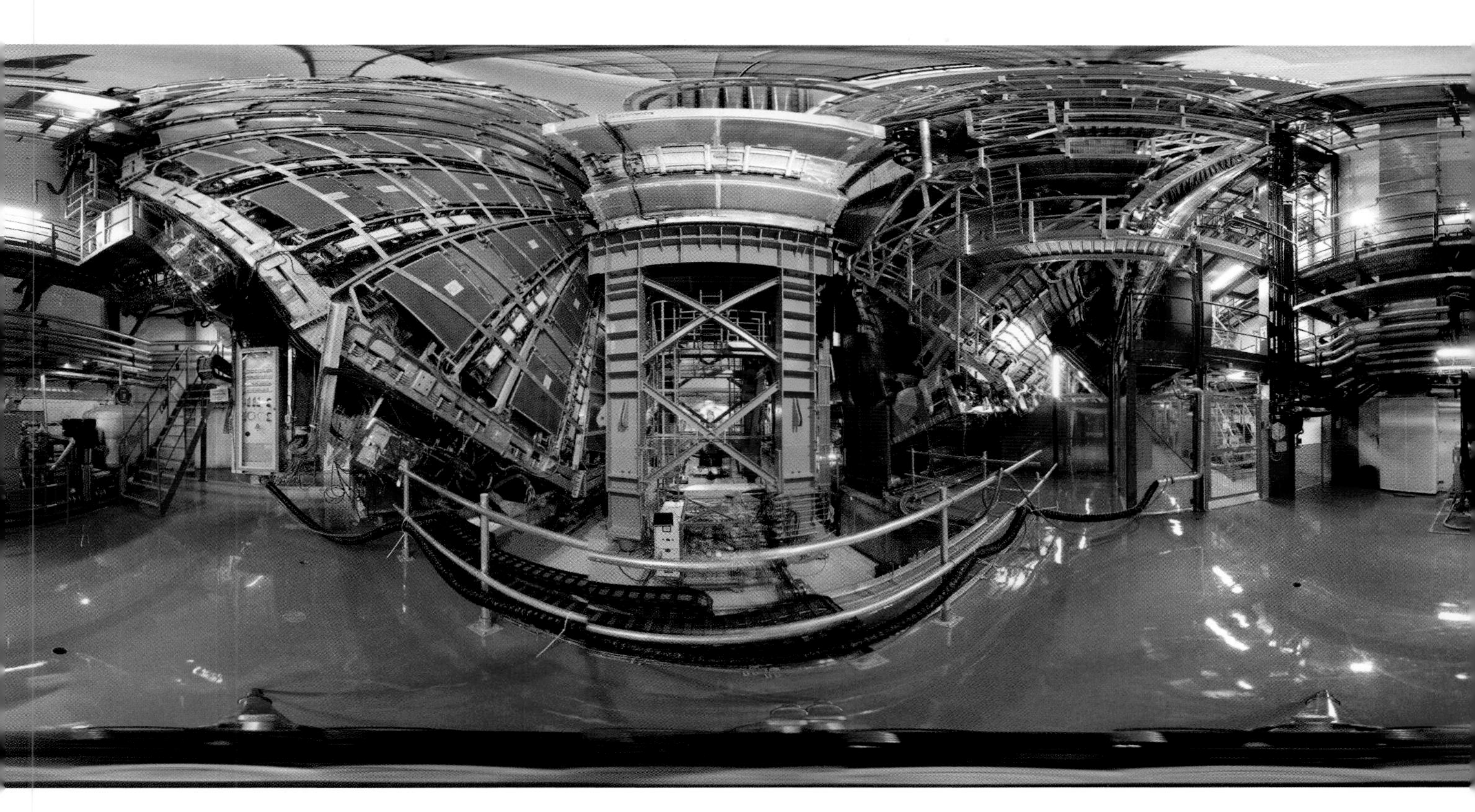

The High Momentum Particle Identification Detector holds the title of being the world's largest RICH detector that's made of caesium iodide, every portion of its 11m^2 area being in action when the Large Hadron Collider gives it some particles to scrutinise.

Measuring the energy of particles are the calorimeters. In general they do this by stopping samples of particles and absorbing their energy. Only muons and neutrinos can pass through calorimeters; everything else is brought to a halt.

ABOVE A 360-degree view of the ATLAS cavern. *(CERN)*

LEFT Installation of a Di-Jet Calorimeter module for measuring the energy of hadrons inside ALICE. *(CERN/Harsh Arora)*

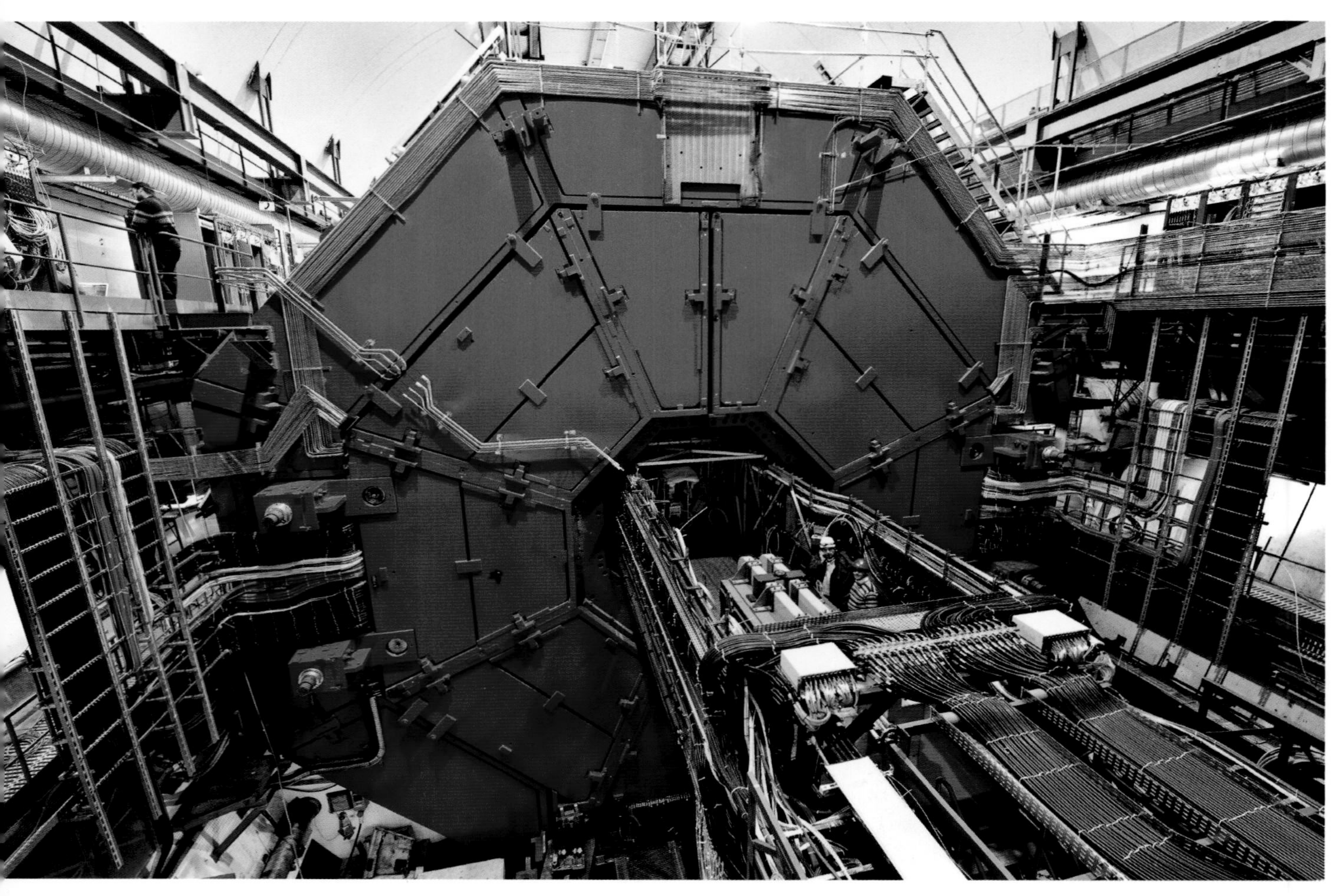

ABOVE A view of ALICE with its magnet door closed. *(CERN/Mona Schweizer)*

Charged particles such as electrons and positrons deposit their energy in the Electromagnetic Calorimeter, or EMCal. This contains 100,000 lead scintillator tiles, distributed among 13,000 'towers' grouped into ten 'super-modules', which in turn are attached to a support frame sandwiched between the ALICE magnet's coil and the Time-Of-Flight detectors. This support frame weighs 20 tonnes on its own, while the entire EMCal weighs in at 100 tonnes and contains 185km of fibre optics. When the particles strike the scintillators and deposit their energy, the lead glows, producing a signal that's read out along the optical fibres, with the intensity of the signal proportional to how much energy was deposited by the particle.

Photons of all energies that are directly produced by the collision are also stopped by a calorimeter, specifically the Photon Spectrometer, or PHOS, comprising large numbers of lead tungstate crystals kept at -25°C (-13°F). These crystals also scintillate, or glow, when impacted by a photon. Both EMCal and PHOS are placed back-to-back within ALICE.

Staying with the topic of photons produced in the beam collisions, ALICE's Photo Multiplicity Detector (PMD) is able to measure how many are formed in collisions and which directions they tend to move in. To avoid contamination by charged particles that could give off false signals, its foremost layer is a detector that filters out or 'vetoes' charged particles, while the photons continue on and meet a second detector where they trigger a signal across several cells, unlike hadrons that activate only one cell upon impact. Instead, the Forward Multiplicity Detector (FMD) is sensitive to charged particles moving out in the forward regions, where particles flung out of the collision at shallow angles to the beam can be detected. The FMD is made from five silicon discs, each with 10,240 detector channels.

What none of the calorimeters or subsequent

ABOVE One of the modules of the ALICE photon spectrometer. There are five modules in total, each containing 3,584 lead-tungstate crystals, which scintillate when high-energy particles pass through. *(CERN/Maximilien Brice)*

RIGHT Technicians insert the last of the lead-tungstate crystals into ALICE's photon spectrometer. *(CERN/Maximilien Brice)*

BELOW A close-up of the lead-tungstate crystals in ALICE's photon spectrometer. *(CERN/Maximilien Brice)*

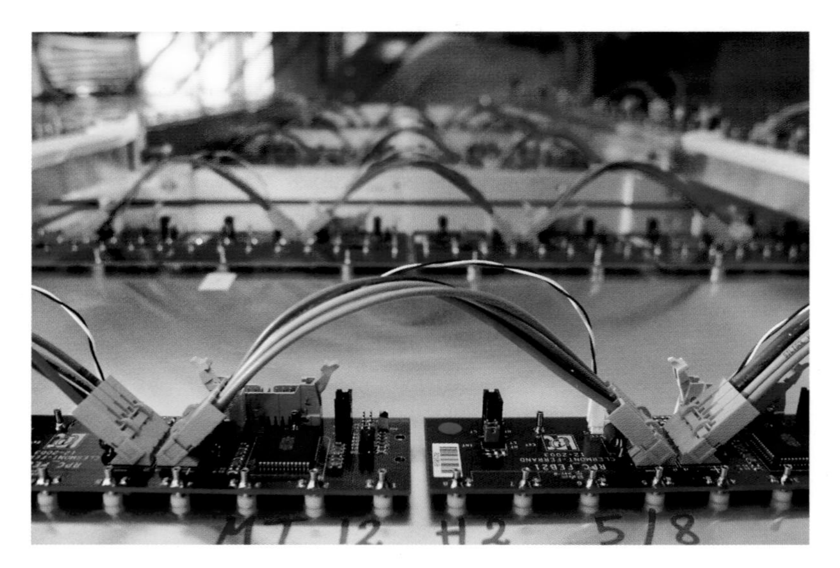

ABOVE Electronics of ALICE's Muon Spectrometer trigger chambers. *(CERN/A. Saba)*

detectors can stop are the muons. For these a specialised instrument is required, namely the Muon Spectrometer, located in the forward regions of the ALICE experiment. The Muon Spectrometer is able to slow and track muons by means of a system involving a thick front absorber and a muon filter made from a wall of iron 1.2m thick. Tracking detectors in the instrument measure the best events.

Detecting the particles is all well and good, but there's one crucial piece of the puzzle that we've neglected thus far, and that's the collision itself. Its energy is recorded in the debris of the collision, and four instruments made from high-density materials – the Zero Degree Calorimeters, the T0 (T-zero) Detector, the V0 (V-zero) Detector and the aforementioned FMD – are able to measure the number of particles produced in the collisions and their overall spatial distribution.

Let's start with the T0 Detector. T-zero refers to the starting point for the collision, *ie* the time at which it occurs, or time-zero, and therefore acts as a trigger for the entire ALICE experiment and provides a reference signal for the Time-Of-Flight detector. The T0 Detector is made from two arrays of Cherenkov counters at opposite sides of the collision point. Each array is made from 12 counters, which are cylinders filled with a quartz radiator through which a particle passes and emits Cherenkov radiation, and a photomultiplier tube that detects the light and turns it into an electrical signal. Meanwhile, the V0 (vertex-zero) detector measures the centre point of the collision between the heavy ions in the beam, and is formed from two arrays of scintillator counters – termed V0-A and V0-C – held 3.5m away from the collision point.

Finally, the Zero Degree Calorimeters (ZDCs) are four small calorimeters positioned 115m from the collision point, along the line of the beam itself (hence an angle of 0° to the beam)

RIGHT Inside ALICE's Muon Spectrometer. *(CERN/Mona Schweizer)*

LEFT A technician inside the barrel of ATLAS' cryostat, which cools liquid argon to –183°C (-297°F) in the calorimeter. The liquid argon cools metal plates that produce showers of secondary particles when a photon, electron or positron passes through the calorimeter. The secondary particles then cause the argon to glow, enabling the energy of the initial particle to be measured. *(CERN/ Claudia Marcelloni)*

and affixed to the outside of the beryllium beam pipe as it moves away from ALICE. The ZDCs measure what are called 'spectator nucleons', in other words protons and neutrons that emerge from the collision. The ZDC detectors are a stack of metal plates sporting grooves filled with quartz fibres, which give rise to their nickname of 'spaghetti calorimeters'. The quartz acts as a radiator that produces Cherenkov light when a relativistic proton or neutron passes through it. Two of the ZDC arrays are designed to detect protons, while the other two identify neutrons.

The last of ALICE's sub detectors is ACORDE, which is short for ALICE Cosmic Rays Detector. This conducts science independent of the ion collisions that take place deep within ALICE. Exterior to the LHC, cosmic rays from space impact molecules in Earth's atmosphere produce showers of energetic secondary particles that move down through the atmosphere and penetrate the surface. ACORDE is able to detect muons in the cosmic-ray showers hitting the top, where they trigger five-dozen scintillator modules on top of ALICE's magnet, meaning that the experiment gives insight not only into what's happening on the smallest scales in particle smash-ups underneath the ground, but also to violent astrophysical cataclysms that are able to accelerate atomic nuclei to a smidgen under the speed of light and to energies that far exceed what even the Large Hadron Collider is capable of. In that sense, the Big Bang machine might only seem like a pretender – the true particle accelerators lie in supernova remnants and black hole accretion discs.

All these sub detectors produce large amounts of data, but the challenge isn't the wealth of data but rather how to handle the two very different operating modes, specifically the small, frequent collisions between protons that produce relatively few particles, and the giant smash-ups between lead ions that produce tens of thousands of secondary particles in one go.

To achieve a balance in how it records these events, the ALICE data pipeline has a bandwidth of 2.5 gigabytes per second and its Mass Storage System has a capacity of

LEFT The connectors and central elements inside ALICE's V0C box. *(CERN/Jean-Yves Grossiord)*

RIGHT Rows of servers in CERN's computing section. *(CERN/Robert Hradil)*

BELOW An overview of CERN's vast computing complex, which handles petabytes of data each year. *(CERN/Robert Hradil)*

1.25 gigabytes per second. Added up over the course of a year this means that ALICE processes over a petabyte (or a thousand trillion, sometimes called a quadrillion) of data per annum.

Perhaps surprisingly, the computers handling this data aren't exotic supercomputers but a network of hundreds of standard PCs loaded with Linux software called DATE, standing for ALICE Data Acquisition and Test Environment. They're there to make sense of what the Large Hadron Collider is showing them.

ABOVE Inside the busy LHC control room. *(CERN/Maximilien Brice/Jacques Fichet)*

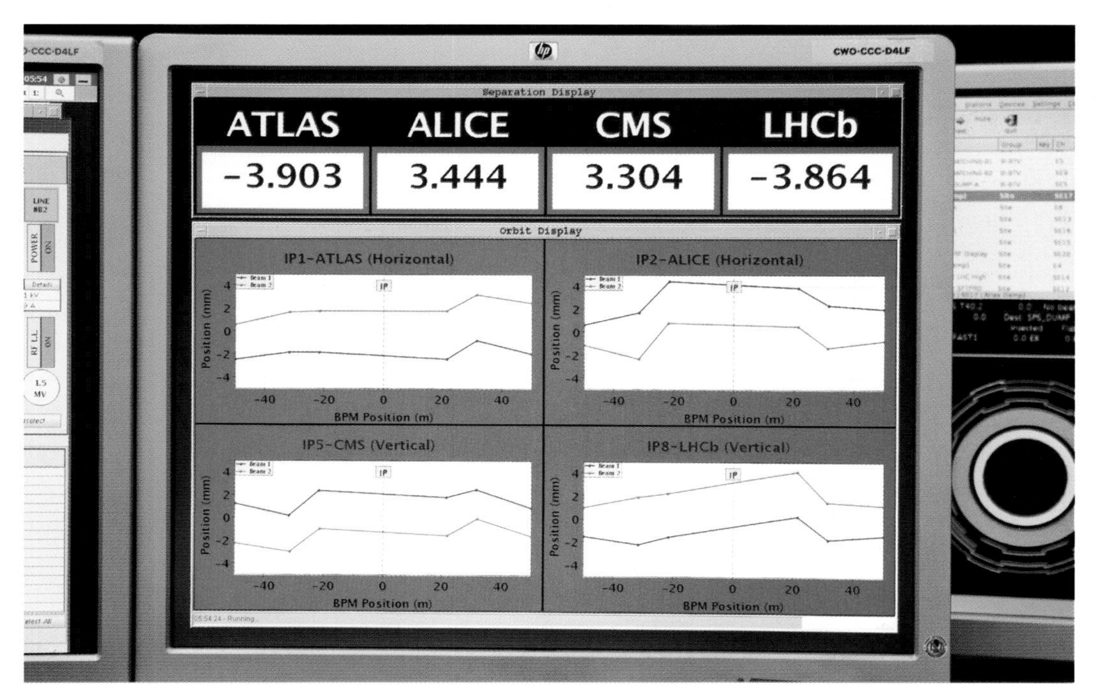

LEFT From the control room technicians are able to monitor the performance of the LHC's four big experiments – ALICE, ATLAS, CMS and LHCb. *(CERN/Maximilien Brice)*

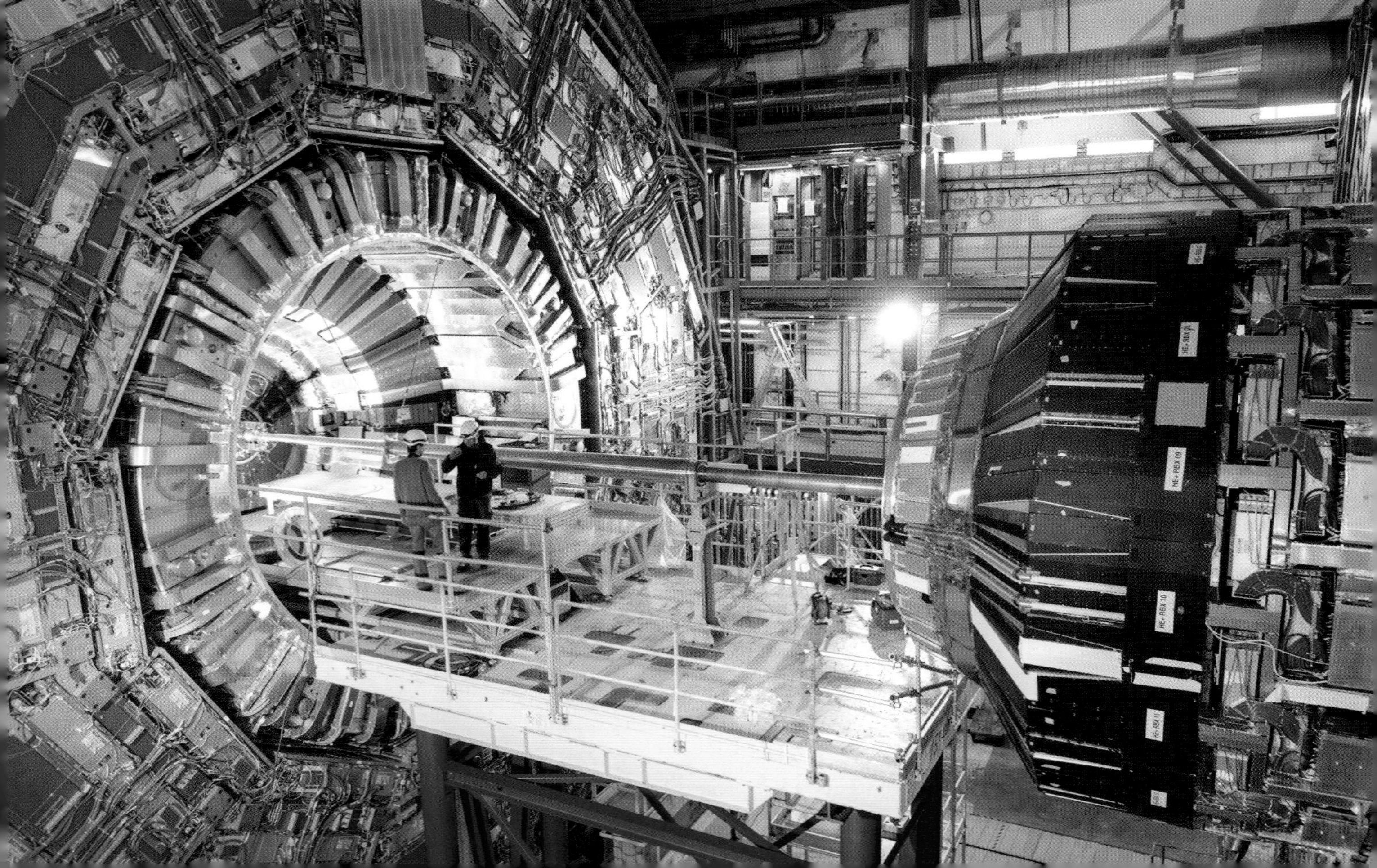
HE+ RBX 09
HE+ RBX 10
HE+ RBX 11

OPPOSITE TOP A view looking down at the CMS, as the Pixel Detector at the heart of the experiment is replaced and upgraded during 2017. *(CERN/Maximilien Brice)*

OPPOSITE BOTTOM The huge scale of the CMS relative to the engineers working on the experiment can be clearly seen here. *(CERN/Maximilien Brice)*

ABOVE Another view of the CMS solenoid magnet. *(CERN/Micheal Hoch)*

CMS: the general-purpose solenoid

Next along the chain is the Compact Muon Solenoid or CMS, which in order to tackle a wide range of physics needs a series of detectors built around its gigantic solenoid magnet. You've probably come across a solenoid at some point in your life – in a car's starter system, for instance, or in an air-conditioning system. The one found at the centre of this experiment is a coil of superconducting cable that smashes out a magnetic field about 100,000 times greater than that around the Earth, or almost 4 teslas. It's composed of several layers of subsystems that facilitate measurement of the momentum and energetic temperament of the aftermath of beam collisions, such as photons, muons, electrons and other such particles.

Let's start from the outer layer of CMS.

BELOW The enormous CMS solenoid magnet, removed from the experiment during the 'Long Shutdown' in 2013, during which the entire LHC underwent upgrades and renovations. *(CERN/Micheal Hoch)*

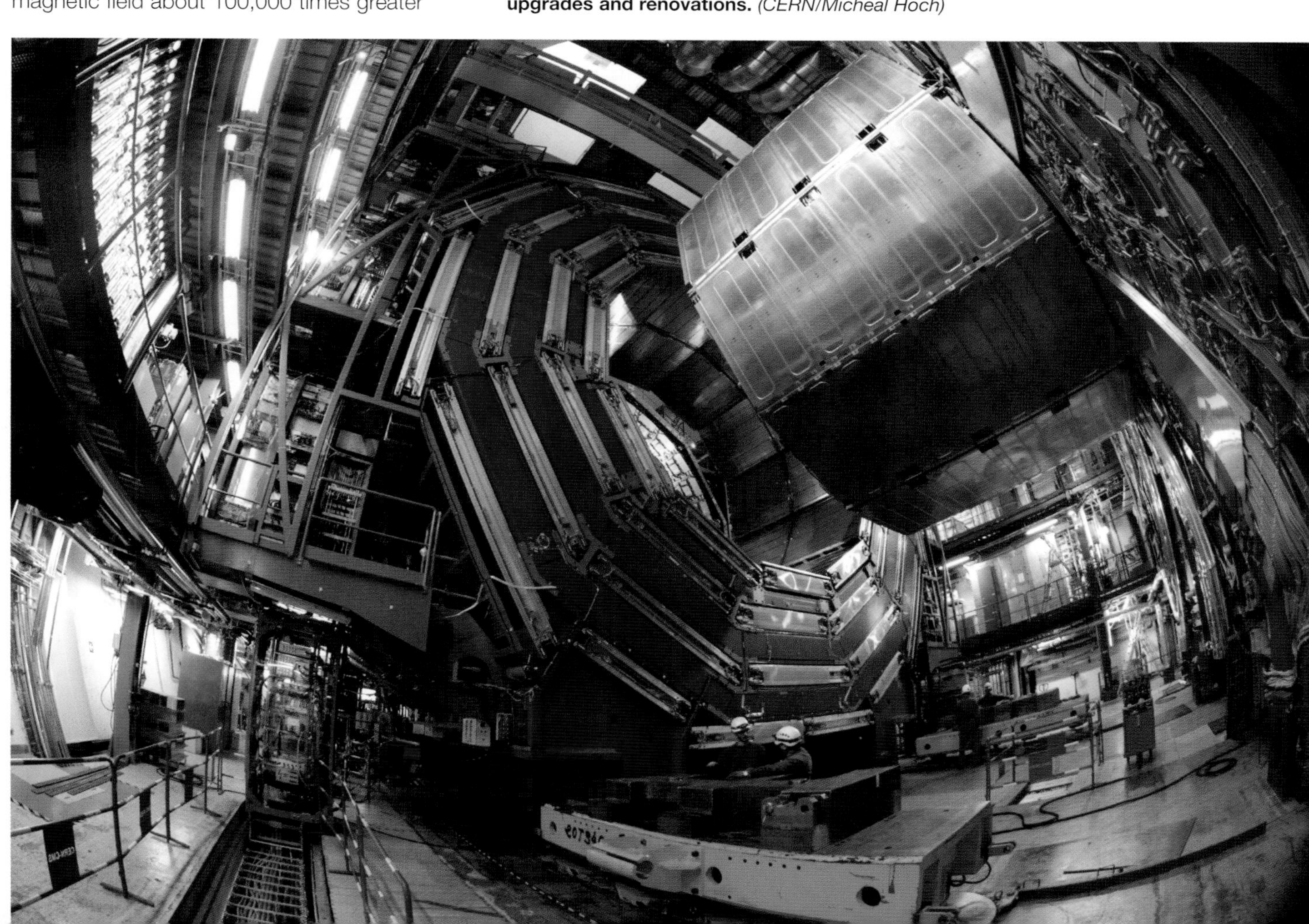

RIGHT The insertion of the beam pipe into the CMS. *(CERN/Michael Hoch)*

This is the tracker, which is able to retrace the particle steps of highly energetic electrons, hadrons and muons, as well as those that erupt from the disintegration of particles that don't live for very long – the beauty or b-quark for example – that's used to look into the discrepancies between matter and 'opposite but equal' antimatter. To be a particle tracker, a device needs to be lightweight. Why? Because if it's too hefty it'll start disturbing the particle beam to the point were it's impossible to reconstruct their path. In fact, CMS' tracker is so featherlight that it's able to take positional measurements within a fraction of the width of a human hair – roughly 10µm. Being the innermost section of the CMS experiment

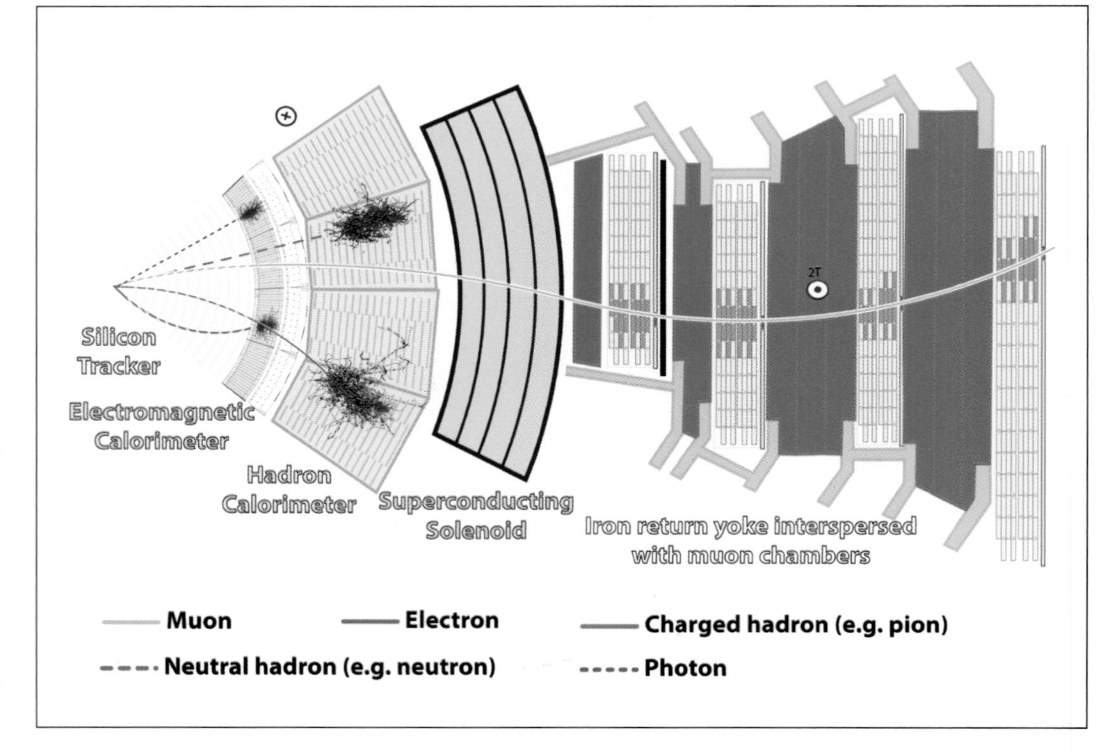

RIGHT A schematic of the CMS experiment, showing the paths of various decay products. The muons travel all the way through to the muon detector at the rear of the experiment. *(CERN)*

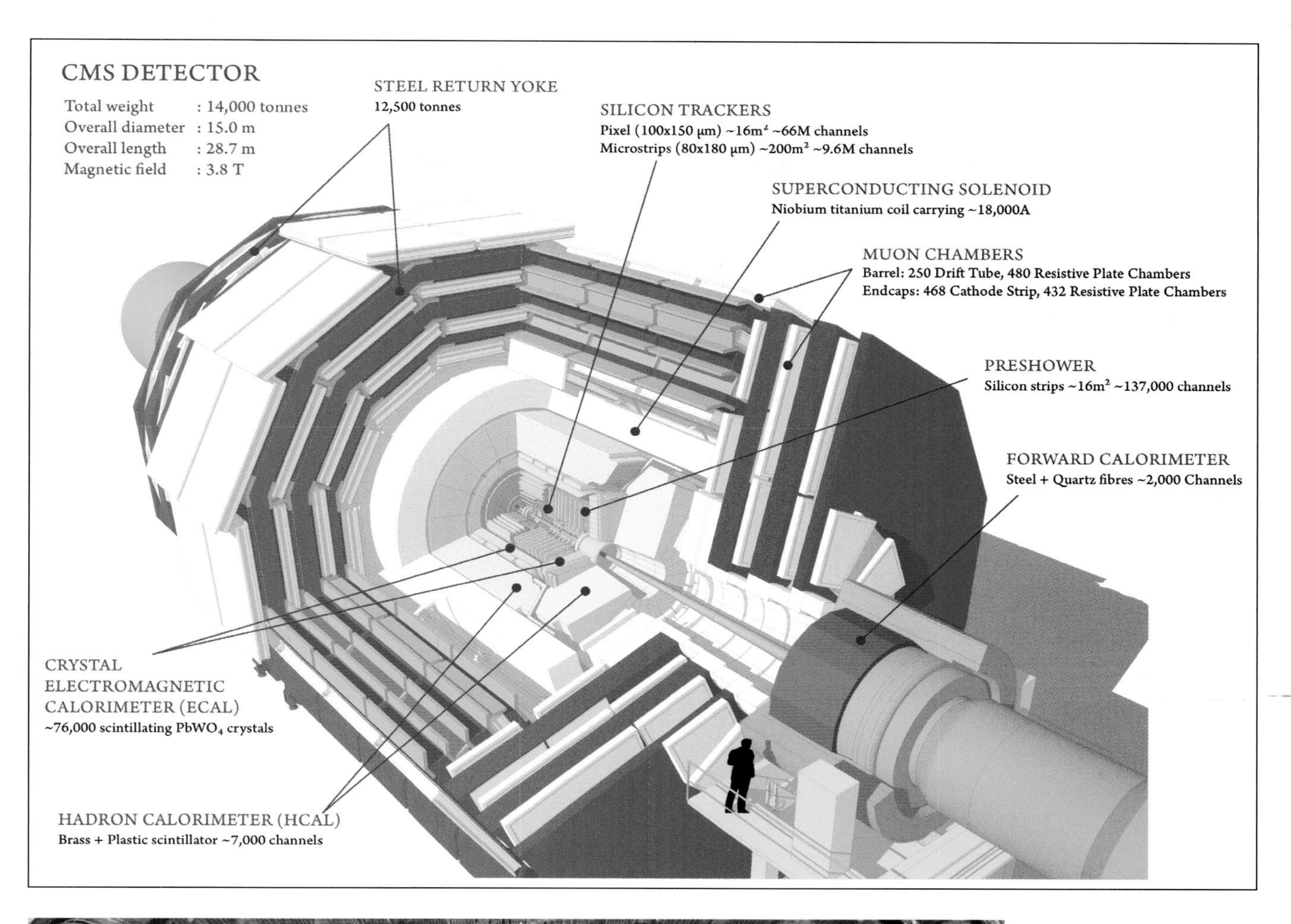

ABOVE An illustrated look at the workings of the CMS and the location of the various sub detectors. *(CERN)*

LEFT The z+ side of the CMS' tracker outer barrel. *(CERN/Maximilien Brice)*

ABOVE The beam pipe running through the heart of the CMS during the replacement of its Pixel Detector. *(CERN/Maximilien Brice)*

means that it's close to where the particle beams crash before new ones spring from the smash-up; and being so close means that it bears the brunt of particles hitting it – that's why this silicon tracker's pixels are located at the centre, to ensure that it catches as many particles as possible. Around this feature sit silicon microstrip detectors, which work

LEFT A close-up of part of the CMS' Pixel Detector during its replacement. *(CERN/Maximilien Brice)*

RIGHT Another close-up of part of the CMS' Pixel Detector, which contains 66 million pixels for tracking the paths of particles. *(CERN/Maximilien Brice)*

CENTRE The CMS' Pixel Detector was replaced during upgrades in 2014. *(CERN/Maximilien Brice)*

in unison with the central pixels to create minuscule electrical signals when particles hit the tracker that are amplified before being seen.

The tracker is no slight feature inside CMS. It's got the area of a tennis court, and is made up of 75 million readout channels, rigged up by some 6,000 connections for every square centimetre. The 66 million pixels that compose the three innermost layers of the tracker are tiny, measuring just 100μm by 150μm. Get into the next four layers and you'll find 9.6 million strip channels, some 10cm in length and 180μm across.

Delve further into CMS and you'll uncover the Electromagnetic Calorimeter, or EMCaL for short. Its main aim is to work out the energy of electrons and photons, in which it's helped massively by its many crystals of lead tungstate, a mineral that often goes by the name of stolzite. Naturally occurring as a brownish-yellow rock, it's engineered to be clear, making it the perfect choice for stopping high-energy particles to get a reading of their energy, with each crystal being 22mm by 22mm in size and 230mm in depth. There's a touch of oxygen in its heavier-than-stainless-steel form, which makes it transparent and helps it work its magic when electrons and photons make their way through. When they 'tunnel' through light is created, in the form of short, obvious flashes of photons that are equal to the energy of the particle ECAL is dealing with.

To look at, ECAL resembles a barrel closed off by two end-caps. It's basically like a bridge between the tracker and the Hadronic Calorimeter, or HCAL for short. In total, ECAL features a whopping 61,200 crystals in 36

RIGHT The lead tungstate crystals that are used as scintillators in the CMS' electromagnetic calorimeter. *(CERN/Laurent Guiraud/Patrice Loïez)*

LEFT Twenty Supercrystals mounted on to one of the D-shaped end-caps of the CMS' Electromagnetic Calorimeter. The crystals detect the energy of photons coming from the particle collision. *(CERN)*

modules carrying 1,700 crystals apiece and weighing in at an impressive three tonnes. The end-caps are studded with a further 15,000 crystals, arranged around detectors that give CMS the power to work out what is a single high-energy photon that's evidence of some exciting physics, and what isn't.

We've briefly touched on HCAL, the next layer of CMS. The clue to what this calorimeter does is in its name: it likes to measure the energy of hadrons, which you'll remember are composed of the gluons and quarks, some examples being neutrons, pions and kaons. It also keeps an eye out for the particles that don't like to interact – the uncharged members of the subatomic family, like neutrinos. HCAL is quite dense. Its bulk material is a mixture of brass and steel (the former of which has been recycled from Russian artillery shells!), while scintillators composed of plastic produce the flashes of light that blast out when a particle makes its way through the transparent tile arrangement. Here, CMS gets its readout with the help of photodiodes, semiconductors that are able to turn the light they receive into electrical current.

Meanwhile, the components of an Hadronic Forward (HF) detector sit just 11m either side of where particle beam crashes into particle beam. The HF is interested in angles, more precisely those created by the subatomic wreckage as they fly in all kinds of directions relative to the particle beam. Since it can get pretty congested at the interaction point, HF likes to use a selection of steel absorbers and fibres fashioned out of quartz for suitable readouts as well as getting an idea of the CMS' luminosity system.

Next up is CMS' big magnet. This is the

LEFT Technicians in front of the CMS' Hadron Calorimeter end-cap prior to full assembly. *(CERN)*

heart of the experiment, around which all its layers are built. To measure the particle's charge-to-mass ratio, it turns into an influencer with its 3.8-tesla magnetic field, gently leading them in a curved path in order to tease information from them. Simply speaking, the more momentum a particle has the less its path is curved by the magnetic field, making it quite easy for CMS to take a reading. If you asked a Large Hadron Collider scientist to tell you how strong CMS' magnet is, they'd tell you that it's packed with an energy equal to two billion joules. That's the equivalent of igniting about half a tonne of TNT. Should CERN ever experience a magnet quench, CMS is ready to handle it – dump circuits (cables that trail from the power converter straight through to the cooling cryostat) carry the energy away.

The final layer of CMS is where you'll find the muon detector and what's known as a return yoke, an iron structure that weaves through the detectors and surrounds the magnet's coils. It reaches for 14m and sifts through the particles it comes across – anything that isn't a muon or a weakly interacting particle, such as a neutrino, isn't allowed to pass through. Muons are able to make their way through layers upon layers of iron without feeling the need to interact. They're also unable to be stopped by most of CMS' calorimeters. Instead, chambers at the end of the experiment are given the job of registering signals. These detectors are found at the very edge of the experiment, which is the only place where CMS can get a signal flavour of muons on their own.

So how does the CMS work out that it's dealing with muons? It uses triple detectors, the first being the drift tubes, which use a barrel

BELOW The Pixel Detector, removed from the core of the CMS to be inspected by technicians. *(CERN)*

LEFT The end-cap belonging to the CMS muon detector. *(CERN/Michael Hoch)*

set-up to measure the positions of the particles. These measure 4cm across and, on the face of it, are made up of a stretched wire within a volume of gas that's able to track the muons as they make their way through, knocking electrons off of the gaseous atoms as they go, following the electric field. The electrons' destination is a positively charged wire and, as they knock into it, they ping out information about how far away the muon is. Each drift tube measures about 2m by 2.5m. Some 12 sheets of aluminium are arranged in three groups of four, each possessing 60 drift tubes. It's the responsibility of the outermost groups to work out one set of coordinates, whilst the middle group knocks together the other parts of the position.

Cathode strip chambers can be found in the end-cap discs of the drift tubes. It's here where the magnetic field isn't so uniform; in fact it's quite uneven and the rate of particles is quite high. CMS' cathode strip chambers are made up of positively charged wires crossed with negatively charged strips of copper. They sit inside a volume of gas, making the presence of muons well and truly known, bashing electrons off the atoms with which they come into contact, which – in turn – start to cluster around the positive anode wire. Meanwhile the atoms, now that they've lost some of their electrons, have become positive ions and float towards the negativity of the copper cathode. At the anode you get an avalanche of electrons, while the gas ions create a pulse of charge through the strips. The wires and strips are perpendicular – or at right angles – to each other, which means CMS is able to determine the position for each particle that passes by. While this is going on, the cathode strip chambers are multi-tasking; seeing as the wires are so closely spaced together, they're quick enough to serve as detectors for triggering and knowing when muons have crossed the six layers found in each of the chambers.

LEFT The drift tubes being installed on the CMS muon detector. *(CERN)*

ABOVE **A completed segment of the CMS muon detector, with drift tubes attached between the lights.** *(CERN)*

LEFT **Inside the busy CMS control room.** *(CERN)*

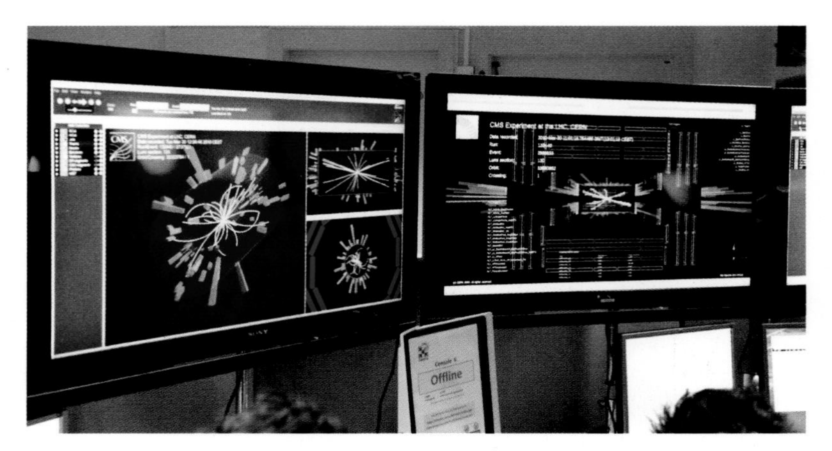

ABOVE Monitors track events during the operation of the CMS. *(CERN/Lussa Tuura)*

BELOW An intimate view of the CMS as it was being opened in 2013 for upgrades to be made. *(CERN/Michael Hoch)*

A fast signal needs to be generated when a muon makes its way through the muon detector. For this, two parallel plates – one positively charged, the other negative – do the job nicely. These are the resistive plate chambers that can be found locked inside the central barrel region and inside the end-caps. Being made of a highly resistive material, spaced out by the gas volume, they produce a similar action to the other chambers inside the confines of the Large Hadron Collider – and that's electrons getting knocked off their atoms. A signal is generated that tells CMS some secrets about the muons, specifically their momentum. The resistive plate chambers also make a lightning-speed decision on whether the data needs to be thrown away or not.

LHCb: subdetectors upon subdetectors

The easiest way to envision LHCb is as a gauntlet of detectors that particles have to run. It's located under the French village of Ferney-Voltaire and is unlike the Large Hadron Collider's other experiments; most notably it's not enclosed. Rather, it's an open chain of various detectors that rest close to the points where the protons smash into each other inside the LHC's beryllium piping. You know from the previous chapter that LHCb's job is to look into the decay of particles that fall apart into the so-called beauty or b quarks and anti-quarks, also known as B-mesons. When these particles are born, they race forward at a shallow angle in the direction of the beam. That was the selling point for engineers – they could make the LHCb more 'open plan'. As a result, each and every detector is stacked behind another like a set of dominoes.

The first detector along the line is located near the particle collision point. It's known as the Vertex Locator, or VELO. The remaining detectors are spread across a length of approximately 21m, ending in the Muon System that's at the far end of the LHCb experiment. As you can imagine, VELO gets a front row seat to the particle-crashing action. It's also the only detector that sits inside the beryllium tube, with 42 silicon sensors each 0.3mm in thickness. The proton beams race through a gap in the middle of the sensors, which are crescent-moon shaped and rest just 7mm from the beam itself, giving them the chance to take data. When they're not in use they like to take a step back, staying out of the way during the beam injection and stabilisation phase. B-mesons are known for being exceedingly unstable, crumbling within just one millionth of a millionth of a second after they're made. That's where VELO has to be quite quick in its detection, sitting close to the beam and sorting the beauty particles from the

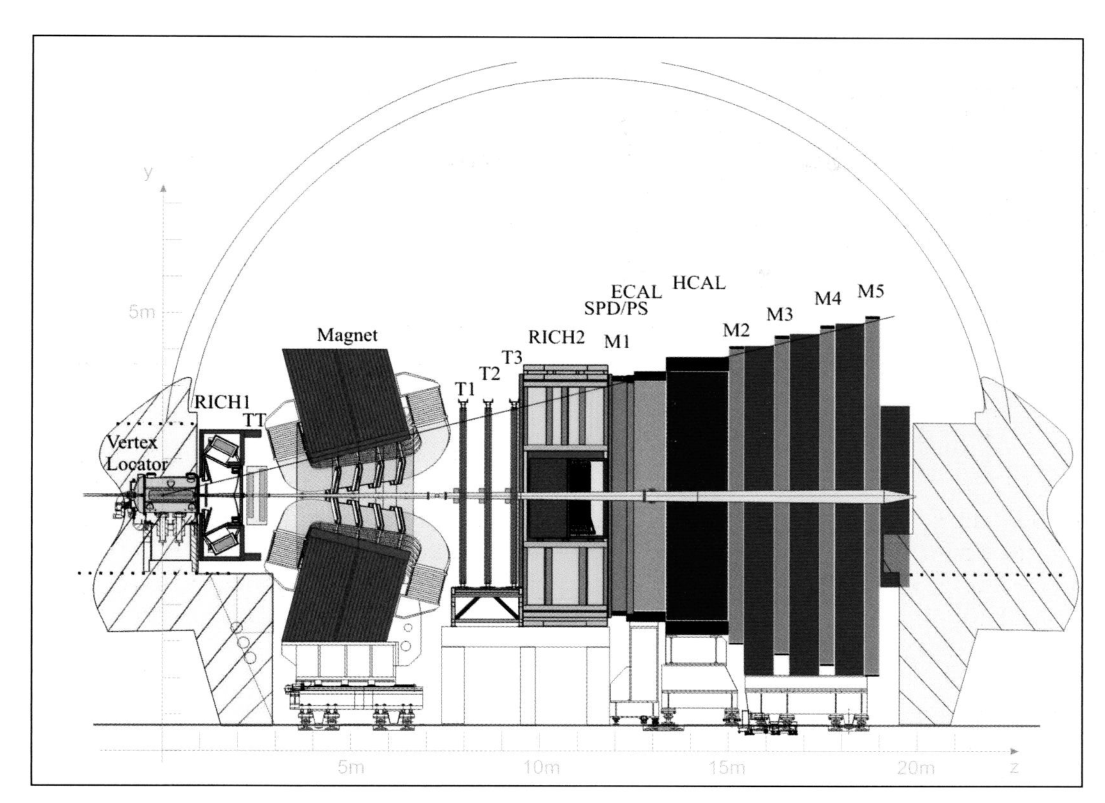

LEFT The layout of the LHCb. *(CERN)*

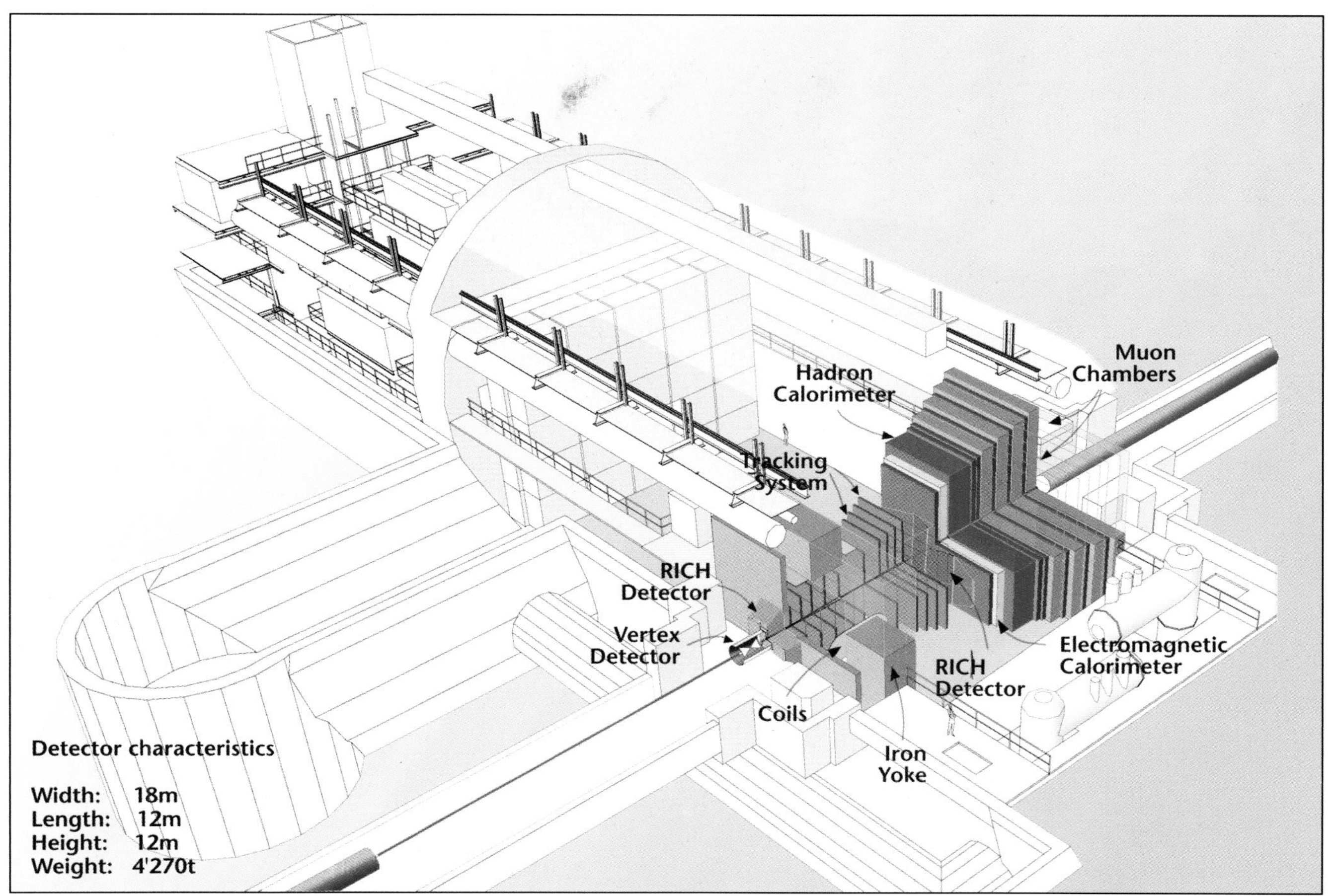

BELOW A diagram showing the layout of the LHCb experiment. *(CERN)*

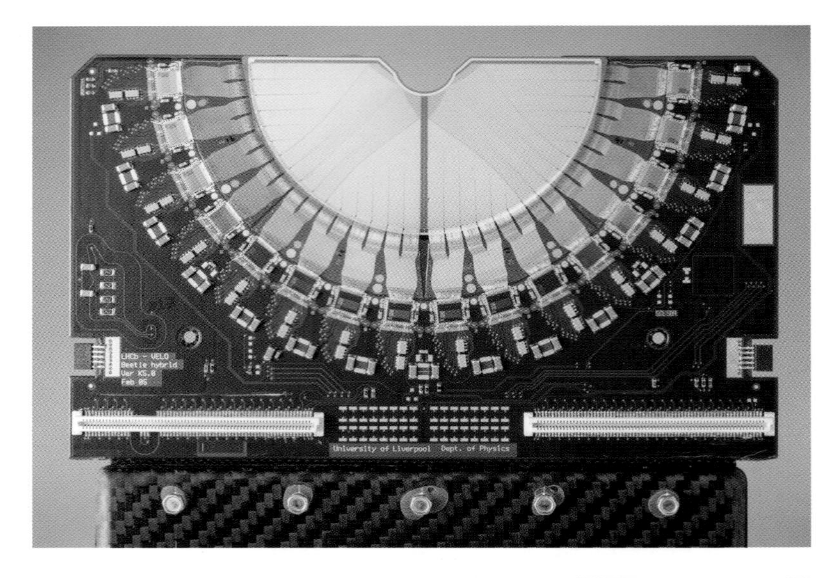

LEFT One of the silicon detectors belonging to the VErtex LOcator (VELO) detector in the LHCb. The silicon detector is the part that looks like a CD cut in half. *(CERN/Maximilien Brice)*

CENTRE Several of the VErtex LOcator's silicon detector modules, which are small enough to fit inside the LHC's beryllium pipeline. It's the first detector inside the LHCb instrument to detect the particles from the immediate aftermath of a particle collision. *(CERN/Maximilien Brice)*

particle beam wreckage. In fact, these particles don't get a chance to travel that far after they're made, but VELO is quite sensitive at pinpointing B-mesons to within 10μ, a micron (μ) being one millionth of a metre, before they disintegrate into their daughter particles.

The B-mesons don't hang around, but the hadrons – kaons, pions and even more protons – into which they fall apart do, at least for a while. That's where the RICH detectors, known simply as RICH-1 and RICH-2, come in. RICH-1 makes its appearance directly after VELO, whilst its accomplice can be found a bit further downstream, after the tracking stations and the experiment's powerful dipole magnet. As well as having similar names, they also have the same function. They're armed with radiators packed with a dense, transparent material, through which light travels much more slowly than through the vacuum, creating the Cherenkov radiation that we've found inside ALICE. Inside this section of the experiment can be found the colourless gas known as perflubatane – a hydrocarbon that you can find inside some fire extinguishers – as can a silica called aerogel, a remarkable substance that's one of the lowest density solids in the world, and doesn't look too dissimilar to solid fog. Both materials are perfect for hunting out those low-momentum particles that spring from the decay of their mesons. Meanwhile, RICH-2 makes use of colourless, non-flammable gas tetrafluoromethane for

LEFT Photomultiplier tubes in the RICH-1 detector in the LHCb experiment. These detect the flashes of Cherenkov light signalling the passage of a particle. *(CERN/R. Plackett)*

seeking out the particles with quite a fierce momentum which have shallower angles to the particle beam from which they originate. You've probably heard of tetrafluoromethane before; it's used in the process of etching silicon wafers in the electronics industry.

Of course, knowing the mass and momentum allows the LHCb team to work out what kind of particle they're dealing with. It's here where the Cherenkov flash is helpful. It's reflected off the mirrors found in both RICH detectors, RICH-1 using mirrors made out of a lightweight carbon-fibre reinforced polymer – more efficient at dealing with the lower-momentum particles – whereas RICH-2, needing to handle higher-momentum particles, makes use of hexagonal glass segments. The sensors that grab this reflected Cherenkov light are known as Hybrid Photon Detectors, HPDs for short; the light occurs where electrons are excited and is released when a photon hits a photocathode. The electrons have a need for speed, accelerated by 20kV on to a silicon detector array packed with 1,024 pixels, $500\mu m^2$ in area and arranged in a 32 by 32 matrix.

The aforementioned tracking detectors are made up of four tracking stations. One of them – dubbed the Silicon Tracker – is separated from the rest, sandwiched between RICH-1 and the experiment's magnet. The three remaining trackers, known simply as T1, T2 and T3, make up the 'Outer Tracker' and sit on the other side of the dipole, just in front of RICH-2. Inside the Silicon Tracker you'll find two detectors, the Trigger Tracker (TT) and the Inner Tracker (IT). The former is the larger of the pair, measuring some 150cm wide and 130cm tall, whereas the IT is a smaller 120cm across and 40cm tall. Both make use of silicon microstrips about 200µ in size. It's here where charged particles crash into silicon atoms, stripping them of their electrons. An electric field of several hundred volts is then applied, which allows charged particles to be collected on one of 270,000 electrodes. The detectors tell us which of these electrodes the electric signal is coming from, revealing the path of the original particle to within 0.05mm. When it comes to materials, silicon is used due to its impressively high spatial resolution, a feature that's great for

ABOVE A clear view of one of the LHCb's Outer Tracker stations as it moved into its final position. *(CERN/R. Lindner)*

BELOW Where the pipeline enters the LHCb Outer Tracker. *(CERN/R. Lindner)*

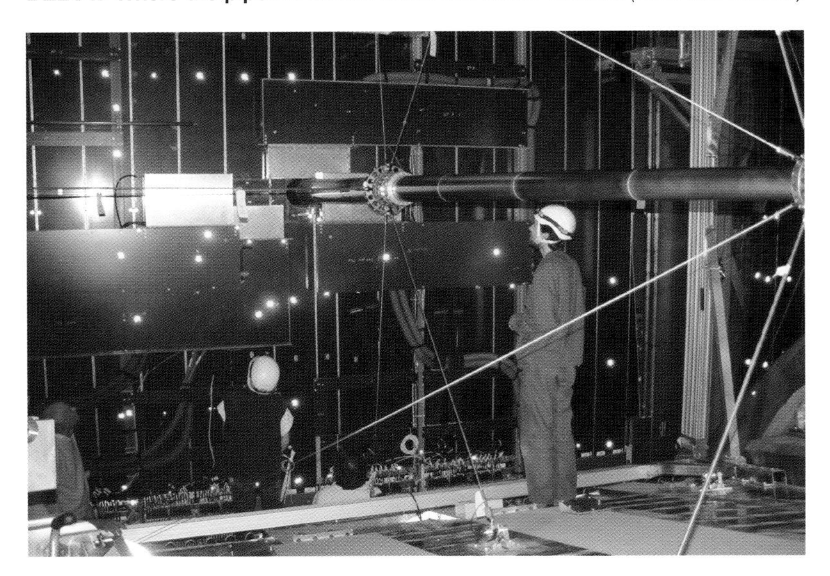

ABOVE The pipeline between the M1 and M2 walls in the LHCb's muon detector. *(CERN/W. Baldini)*

working out which particles are which as they pass through the detector.

Silicon microstrips are considered to be expensive, so it's not possible to make a great deal of the LHCb's trackers out of this metal. The three detectors that make up the Outer Tracker have instead been manufactured out of straw-drift chambers filled with mainly argon and a fraction of carbon dioxide, covering an area of about 6m by 4.9m. As usual, it's set up in such a way that when a charged particle makes its way through one of the chambers – each roughly 5mm in diameter – it ionises the gas, liberating electrons that then race over to the anode wire found at the centre of each tube. The time it takes for the electrons to impact the anode reveals the trajectory of the particle as it makes its way through the set-up, and piecing this data together from each and every tube means that LHCb is sure of its entire path through the tracking station.

Dissect an Outer Tracker station and you'll find that it's made up of 72 modules, each with up to 256 straw tubes, made out of foil and polyimide film, which is produced from Kapton doped with carbon and acts as a negatively charged cathode. On the outer layer you'll find a polyimide-aluminium laminate, capable of shielding the tubes, which combines with

BELOW The beam pipe passing through the LHCb. The particles pass from right to left. In the centre are the magnets, and on the left is the outer tracker station. *(CERN/Maximilien Brice/Mona Schweizer)*

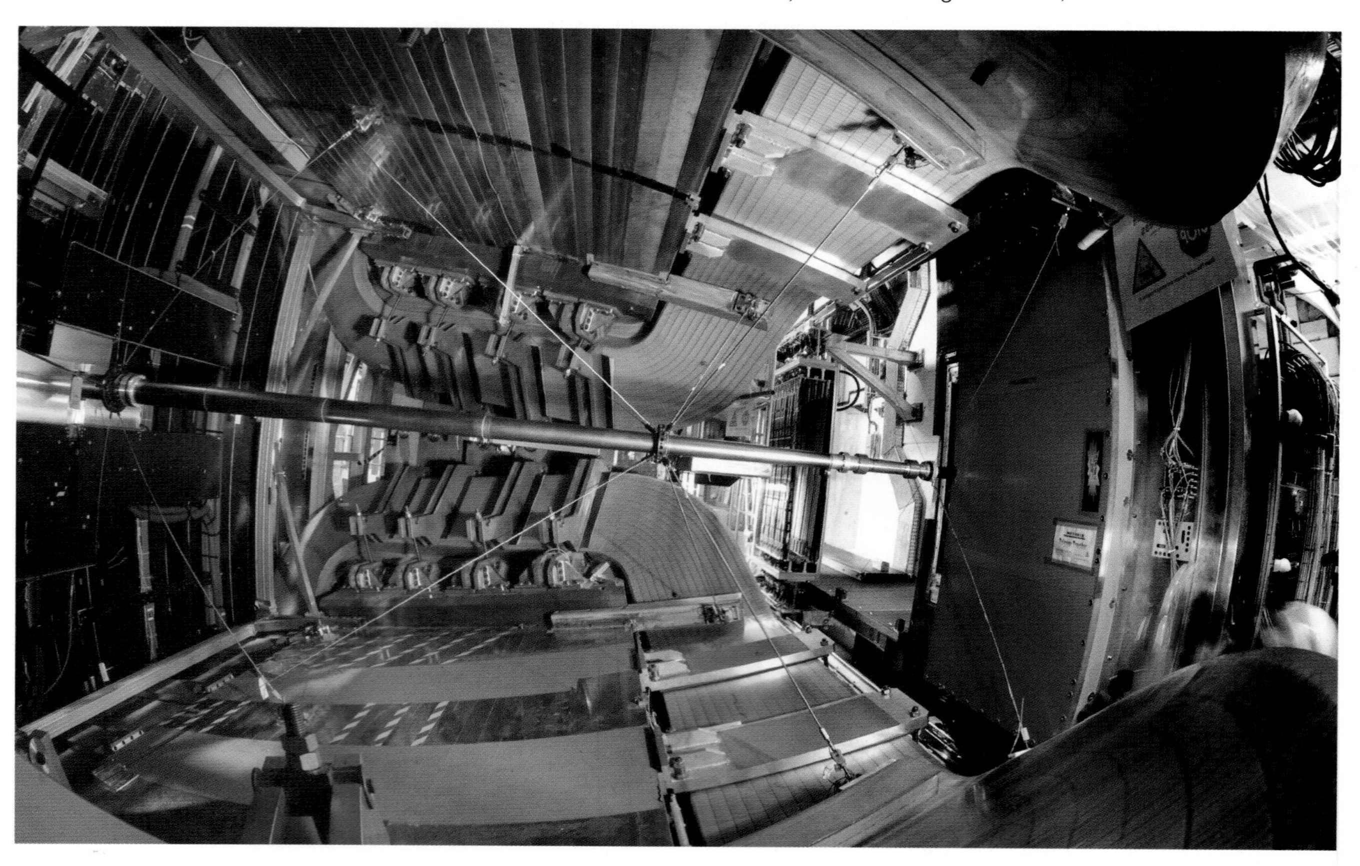

the anode wire to transfer detection signals to the front-end electronics. The tubes are all mounted on moving aluminium frames that carry 18 modules each. Inside the modules, the straw-tubes are arranged so that they sit in two staggered layers, each one tube thick.

Let's not forget the dipole magnet. Acting to curve the paths of charged particles, it's quite a huge affair that's formed from two separate coils. Each weighs in at 27 tonnes, mounted upon 1,450-tonne steel frames. Both of the coils are 7.5m long, 4.6m wide and 2.5m high and are made of aluminium cable that's been wound up exceedingly tight.

Head to the back of the LHCb and you'll find two more instruments, the Muon System as well as two calorimeters. As you'll remember, muons are some of the last decay products to emerge from the collisions between protons and the subsequent decay of B-mesons. Five stations make up the Muon System, simply labelled M1 through to M5, which increase in size the further you move towards the back of the experiment. Add up the detection area of the LHCb and you'll find that it's over 430m^2, a size that's ideal for catching any stray particles produced in collisions. Pumped into each of the stations are 1,380 chambers filled with a concoction of argon, carbon dioxide and tetrafluoromethane. It's here where the muons feel at their most reactive and strip the atoms of their electrons, which are then snapped up by the wire electrodes. The kind of muon that was detected is plain for LHCb to see – a readout signal for one of the 126,000 readout channels that feed into the electronics.

M2 to M5 are right at the back of the LHCb, but sandwiched between M1 and M2 are the various calorimeters. These have two roles. One is to detect and measure the energy of neutral particles, such as neutrons and photons, while the other is to provide information to the trigger system. The entire calorimeter system sports four different layers. The first, directly behind M1, is the combined Scintillating Pad Detector (SPD) and Pre-Shower Detector (PS), each 15mm thick. The SPD determines whether the incident particles have an electric charge or are neutral, whereas the PS determines what type of charged or neutral particle they are. Next up

ABOVE Fibre optic readouts that form part of the LHCb's calorimeter. *(CERN/R. Kristic)*

is the electromagnetic calorimeter (ECAL), followed by the hadron calorimeter (HCAL). ECAL measures less massive particles such as electrons, whereas HCAL detects more massive particles such as protons, neutrons and other hadrons. Both ECAL and HCAL have a structure of alternating layers of metal and plastic plates. When particles strike the metal plates they produce a shower of secondary particles, some of which excite polystyrene molecules in the plastic plates. When excited these molecules emit ultraviolet light, the amount of which is proportional to the energy of the particles that first struck the metal plates.

Last but not least is HeRSCHeL, the High Rapidity Shower Counters for LHCb. It's not actually located in the LHCb experiment cavern, but in the main tunnel itself, just outside the beryllium pipe containing the beam. It's designed to catch protons that don't collide but merely receive glancing blows that cause them to deviate from the beam by just a tiny amount, not enough to be detected by the LHCb experiment, and re-emerge in the beam further down the tunnel, where HeRSCHeL is located. This happens to the majority of particle events, so really the LHCb is interested in events that HeRSCHeL cannot detect, as a way of filtering events that matter to the LHCb from those that don't. HeRSCHeL itself isn't a big piece of equipment, consisting of 20 square plastic scintillators just 30cm in diameter, which produce flashes of light when a proton passes through them.

OPPOSITE One of ATLAS' mighty toroid magnets. *(CERN/Maximilien Brice)*

ATLAS: the world's largest particle smasher experiment

Next along the circuit is the monstrous ATLAS. By volume, this experiment is the largest detector ever constructed for a particle collider experiment. A cylindrical mass of sensors, magnets and electronics, it is 46m long and 23m wide, weighing in at a whopping 7,000 tonnes, equivalent to the mass of the Eiffel Tower. Inside it is 3,000km of cabling. The sub detectors within it are barrel-shaped, concentrically wrapped around the beam pipe like the layers of an onion. Protons collide within the heart of ATLAS at 0.000001% of the speed of light, a billion interactions taking place every second, producing 60 million megabytes of data each second. Of those billion interactions perhaps only a few hundred are of ultimate interest to the ATLAS science team. These are recognised by its triggering system, which selects the events interesting enough for further analysis.

The four primary components of ATLAS are the Inner Detector, the Calorimeter, the Muon Spectrometer and the Magnet System. As its name suggests, the Inner Detector is closest to the collision point in the beam and is the first to detect the decay products from interactions. Extending from mere centimetres from the beam out to a radius of 1.2m, and 6.2m in length, the Inner Detector tracks the trajectory and in particular the momentum of charged particles emanating from the proton collisions, by measuring how the particles interact with various

BELOW ATLAS' subdetectors sit at the very heart, close to the beam pipe where collisions occur. *(CERN/Maximilien Brice)*

ABOVE Technicians in the process of assembling ATLAS' Semi-Conductor Tracker (SCT). *(CERN)*

ABOVE Technicians make the final adjustments to one of the modules of the Transition Radiation Tracker, which is part of ATLAS' Inner Detector. *(CERN)*

RIGHT The view inside ATLAS' semiconductor barrel. *(CERN/Maximilien Brice)*

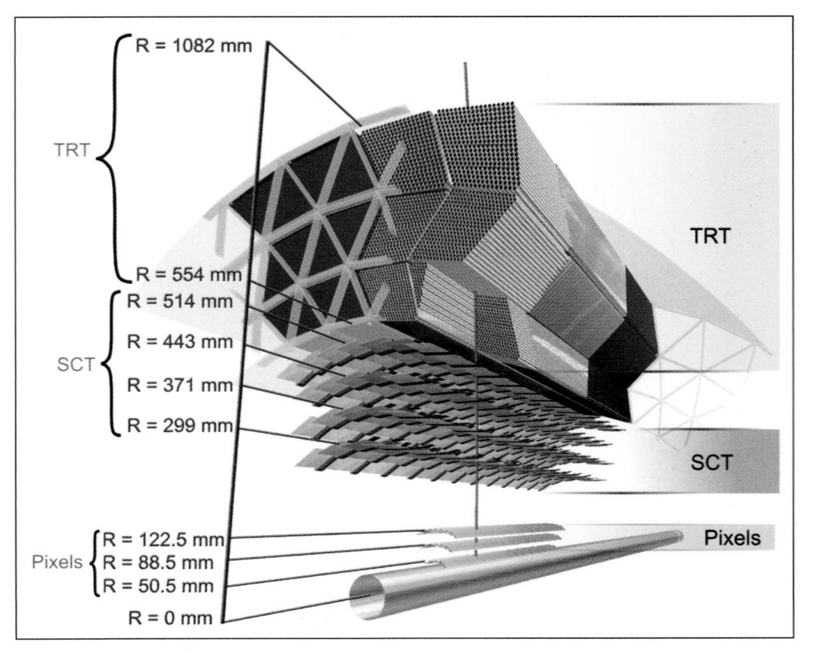

LEFT An artwork showing a cross-section of ATLAS' inner detector, with the radii of the various subdetectors from the beampipe. *(CERN)*

sensors inside the Inner Detector. It consists of three different sub-systems, namely the Pixel Detector, the Semi-Conductor Tracker (SCT) and the Transition Radiation Tracker (TRT).

The Pixel Detector, which is closest to the beam, is made from three cylindrical layers at radii of 5cm, 9cm and 12cm from the beam collision point, filled in total by 1,456 modules of readout chips covered in 50μ by 400μ pixels. There are also three discs on either end-cap, totalling another 288 modules. In total, the Pixel Detector harbours 80 million pixels, corresponding to 80 million readout channels, providing high resolution. The modules, which have dimensions of 6.2cm by 2.1cm, overlap to provide seamless coverage. Each module contains 16 readout chips, with each chip containing 16 by 160 pixels, resulting in 46,080 pixels per module. Being so close to the proton beam, the readout chips are radiation-hardened to withstand 300,000Gy (almost 4,000 times

RIGHT Technicians work inside the ATLAS detector. *(CERN/Claudia Marcelloni)*

the radiation dose required to fight cancer cells) of ionised radiation, and 500 trillion neutrons per square centimetre over a decade of operation.

Next out from the beam is the SCT, which has four double-layers of silicon microstrip sensors instead of pixels, at radii of 30cm, 37.3cm, 44.7cm and 52cm from the beam. The strips are long and thin, just 80μ wide but 12cm long, while the end-caps sport tapered strips. It is able to measure particles in the expanding shower of decay products over a wider area – about 61m² – than the Pixel Detector, but its resolution is not as high, with 6.3 million readout channels.

Finally in the Inner Detector is the TRT, which uses 298,000 straw-tube detectors similar to the Outer Tracker in the LHCb. Each tube is 4mm in diameter, 144cm long and contains a 0.03mm diameter gold-plated tungsten wire and a mix of gases composed of 70% xenon, 27% carbon dioxide and 3% molecular oxygen. Earlier science runs used an alternative concoction of 70% xenon, 20% carbon dioxide and 10% tetrafluoromethane. When a charged particle passes through the tubes it ionises the mix of gases, and an applied voltage of 1,500V accelerates the resulting free electrons towards the gold and tungsten mix wire, producing a current that's converted into a signal. The tubes that are ionised trace the path of the particle. To aid in detecting the particles, materials between the straw tubes with a high refractive index instigate transition radiation. It's here where the radiation is emitted when a fast-moving particle passes through the transition between different materials, which can produce a much stronger signal in some of the straws when the radiation interacts with the xenon gas. This enables the lightest-charged particles, such as electrons, to be distinguished, because the transition radiation is stronger for particles moving at closer to the speed of light, which usually corresponds to lighter particles.

RIGHT Looking down on the detector in the ATLAS cavern. *(CERN/Claudia Marcelloni)*

ABOVE Inside ATLAS' chamber as it appeared in March 2008. *(CERN/Mona Schweizer)*

BELOW The Central Trigger Processor, which collects trigger information from ATLAS' calorimeter and muon detector. *(CERN/Matteo Franchini)*

The Inner Detector is surrounded by a central solenoid magnet. Its 2-tesla magnetic field, generated by 9km of superconducting wire, is able to bend the trajectory of charged particles, and the degree of bending as measured by the track detected by the Inner Detector gives away the momentum of those particles.

Meanwhile, the second significant component of ATLAS is the calorimeter, which is wrapped around the outside of the solenoid magnet. Whereas the Inner Detector measures momentum, the calorimeter measures a particle's energy. As a particle passes through the calorimeter it loses its energy and is stopped, depositing that energy in the calorimeter's metal plate absorbers, which are made from lead and stainless steel – two dense metals. Rather than all decay products being measured, the particle shower is sampled by the calorimeter system, otherwise nothing would get past it other than muons and neutrinos.

The calorimeter is split into two. Innermost is the electromagnetic calorimeter, which absorbs charged particles and photons – particles that

ABOVE Inside the ATLAS calorimeter before it was closed in 2008 before the LHC went into operation. *(CERN/Claudia Marcelloni)*

interact with the electromagnetic force. The outer calorimeter is the hadronic calorimeter, which absorbs particles such as neutrons that pass through the electromagnetic calorimeter but which interact more with the strong nuclear force. Between the metal plates are sensing media. Associated with the electromagnetic calorimeter is liquid argon, chilled to -183°C (-297°F), which can be ionised by charged particles; the resulting free electrons produce an electrical signal. Meanwhile, the outer hadronic calorimeter is made of only steel plates as opposed to the lead and steel of the electromagnetic calorimeter, and its sensing apparatus is the 20,000kg 'tile calorimeter', which is a huge barrel split into 64 5.6m-long wedges upon which 500,000 scintillating plastic tiles are fixed. When one of these tiles receives the absorbed energy of a particle, it scintillates, emitting light that's then converted into an electronic signal.

As its name suggests, the Muon Spectrometer measures muons, which as we've seen the calorimeters are transparent to.

BELOW A view of ATLAS' calorimeters from below. *(CERN/Claudia Marcelloni)*

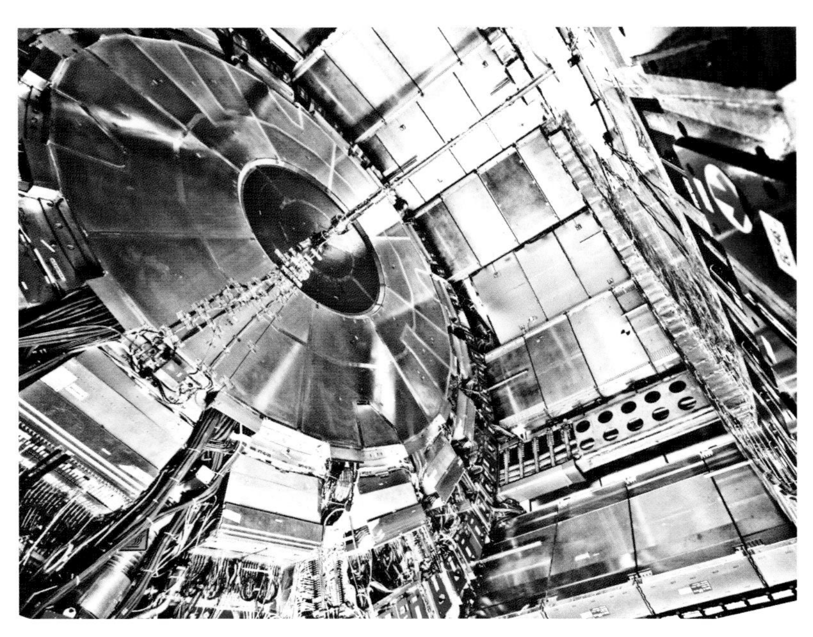

RIGHT An engineer makes improvements to ATLAS' Muon Spectrometer by installing a final layer of chambers. *(CERN/ Claudia Marcelloni)*

FAR RIGHT A view of ATLAS' end-cap magnet toroid. *(CERN/ Claudia Marcelloni)*

OPPOSITE A view of the rear of ATLAS' end-cap magnet toroid. *(CERN/Claudia Marcelloni)*

The Muon Spectrometer is huge, with a total detecting area of $12,000m^2$, and forms the bulk of the ATLAS' volume. It starts at a radii of 4.25m from the beam, on the outside of the calorimeters, and extends out to 11m, which is the full extent of the large detector.

A magnetic field within the Muon Spectrometer is generated by the toroidal superconducting magnets, familiar to readers as the octagonal structure with eight radial pipes housing the magnetic coils used to generate a field of 3.9 teslas. ATLAS also has two end-cap toroidal magnets that can achieve a maximum field strength of 4.1 teslas. To maintain their superconducting properties, the magnets are cooled to just 4.7° above absolute zero.

BELOW The ATLAS Pixel Detector is the innermost part of the experiment's Inner Detector. *(CERN/ Claudia Marcelloni)*

Alongside the magnets is a cornucopia of different detector chambers. The magnets create a field that's able to cause the trajectories of the muons to bend, depending on their momentum. Thin-gap chambers with 440,000 channels measure a muon particle's non-curving coordinate at the ends of the spectrometer, as well as activating the trigger system. The same goes for the collection of resistive plate chambers with 380,000 channels that measure the same non-curving coordinate but in the centre of the detector. Cathode strip chambers have 70,000 channels to more precisely measure coordinates at the ends of the detector.

Meanwhile, there are 1,171 chambers containing 354,240 drift tubes, which aren't too dissimilar to the straw-tubes of the Inner Detector, only larger at 3cm in diameter and up to 6.5m long, which track the curved paths of the muons that have energies of up to 1TeV.

Of course, with all these billions of particle interactions something has to coordinate ATLAS' sensors. This is the aforementioned

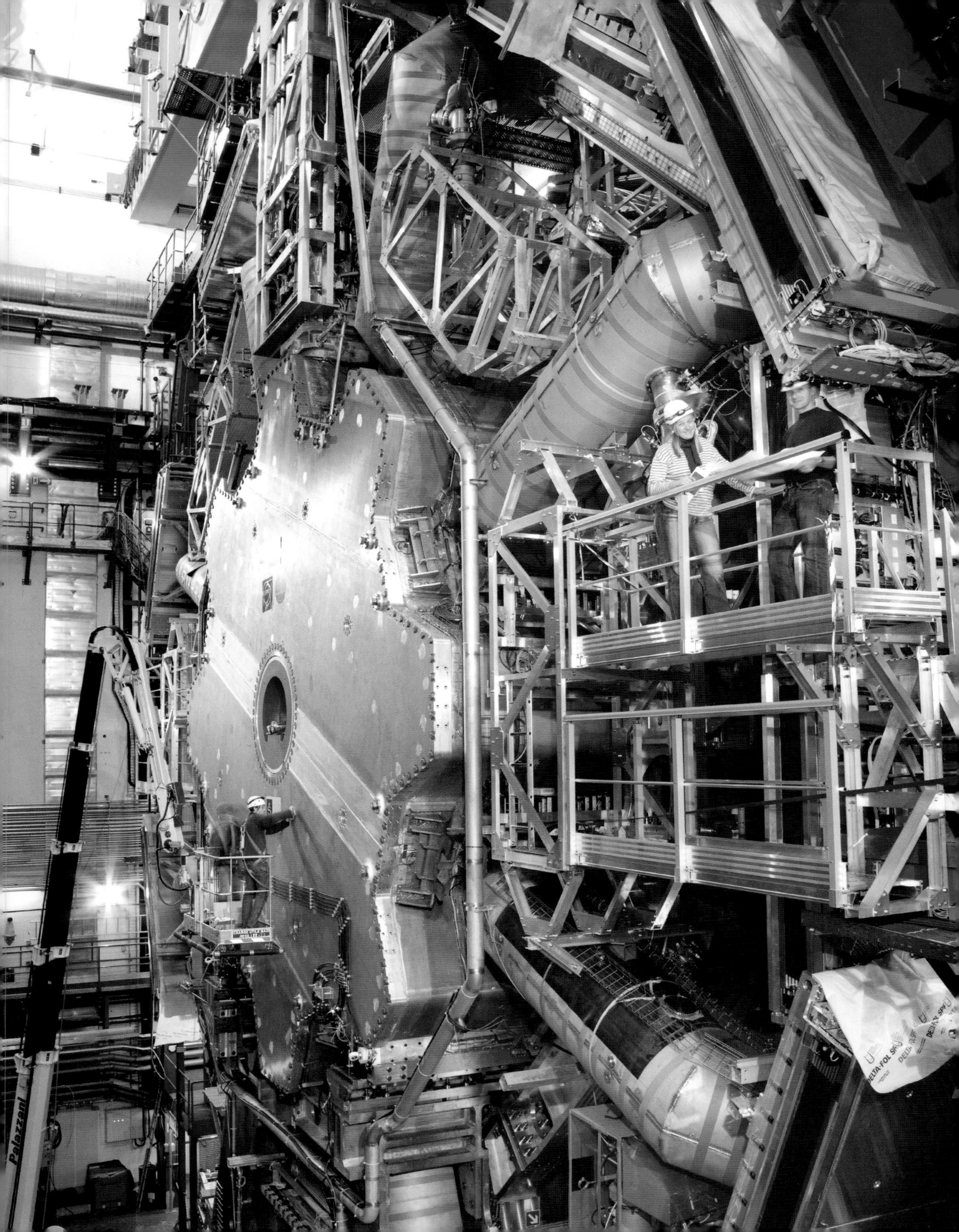
Palazzani
DELTA-FOL SPU

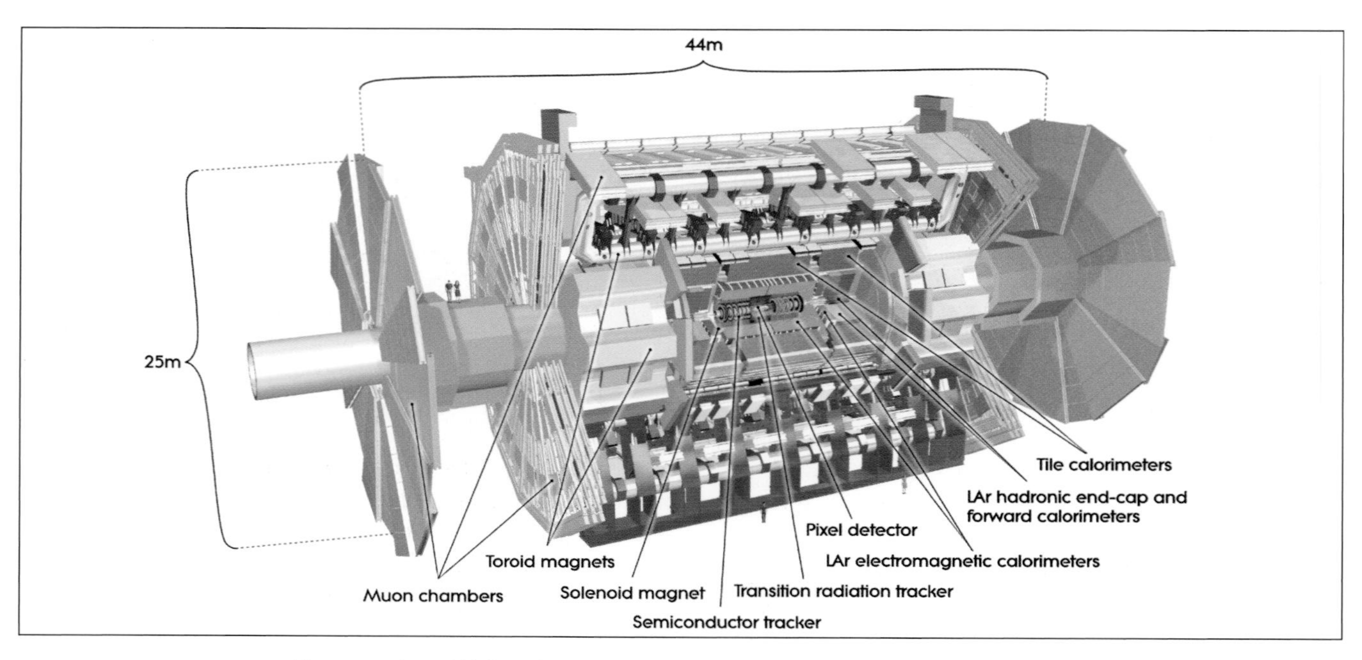

ABOVE A labelled overview of the inside of the ATLAS experiment, showing its various subdetectors and magnets. *(CERN)*

trigger system. We've already seen how the chambers in the Muon Spectrometer trigger the experiment; that's the Level 1 trigger, and also involves data from the calorimeter, which senses the other particles. When an event is triggered at Level 1, the information goes to readout buffers and the system has just two-millionths of a second to decide whether to keep that data. A total of 100,000 events are saved by the Level 1 trigger every second.

However, the data has to now pass through two more filters, the first of which is the Level 2 trigger. Computer processors analyse in greater detail the 100,000 events saved by Level 1 of which only a few thousand are deemed sufficiently relevant to pass through to Level 3. During Level 2 all the various pieces of data from ATLAS' subdetectors are collected into a single memory file, known as 'Event Filter processing', which is conducted by farms of computer processors. Finally, at Level 3 even more computer power is applied, narrowing down the number of interesting events to 200 per second, which are then further scrutinised offline by the ATLAS science team.

The Large Hadron Collider's sentry

As you know by now, the looping tunnel of the Large Hadron Collider is an impressive 27km long, which is a huge length to monitor constantly for faults or damage. Fortunately for the workers at the LHC, their legs are saved from endlessly wandering the tunnel by a roving robot, called TIM (Train Inspection Monorail unit).

There's actually two TIMs at the particle smasher, which wander the entire length of the tunnel by hanging from a monorail attached to the ceiling. The monorail actually predates the Large Hadron Collider, having been built during the days of the Large Electron-Positron collider, when it was designed to ferry workers and equipment around.

In a way the TIM units are like trains, made from linked modules that ride the monorail. The head of the train watches everything around it with its electronic eyes – a pan-tilt zoom camera, a high-definition camera and a thermal infrared camera, all for checking that the structure of the tunnel remains sound. The carriages that the TIMs pull around like beasts of burden are armed with all kinds of gizmos: sensors to measure oxygen levels, to probe the communication bandwidth in different parts of the tunnel or to sniff out radiation leaks. If there's a problem in the tunnel, a TIM will find it.

When they're not working, the twin TIMs spend their time parked up in the CMS bypass tunnel. When called into action, a little 'cat flap' opens in doors, and gaps are cut into fences to allow them to pass unhindered into the main tunnel, which they can traverse at up to 6kph.

The brain inside each TIM is an onboard

Siemens S7-300 PLC – a 'programmable logic controller', which is a computer specifically designed to operate in challenging environments, such as radiation leaks or tunnel fires. The onboard cameras and sensors are all directly linked to the PLC, which processes the data, controls the DC motor and can even perform autonomous tasks that have been programmed into it. The data is then stored on a Siemens IPC or industrial PC. A graphical user interface (an HMI, or human-machine interface) allows engineers on the surface to control the TIM units. The HMI was developed using LabVIEW, or Laboratory Virtual Instrument Engineering Workbench, which is a software platform developed by National Instruments for use with a visual programming language – in other words, a computer program that's written with graphics rather than traditional text code, providing a more user-friendly interface.

The TIMs are never alone – a 4G universal mobile telecommunications service present along the tunnel means they're in constant contact with engineers on the surface or working in the tunnels. All the data they collect and store is then transferred wirelessly back to a central database, where it's published on an internal website for analysis.

The need for TIM was originally raised back in 2007 by engineers both in CERN's engineering department and in the radioprotection group that's part of CERN's health, safety and environment department. The idea of using the robots is to minimise the risk to human workers: the TIMs can find a radiation leak or detect potentially harmful anomalies in the oxygen levels in the tunnel before any humans encounter them. The health of the TIMs is also paramount – they're regularly serviced when the Large Hadron Collider is inactive and their sensors and thermal camera are calibrated every two years. They'll also get help soon; more TIM units are planned, which will cut down response time to problems in the tunnel and increase tunnel coverage, making the LHC safer for everyone – whether they're humans or robots!

ABOVE **One of the Train Inspection Monorail units (TIMs) that travel around the LHC, inspecting and detecting faults.** *(CERN/Alessandro Masi)*

BELOW **TIM is armed with infrared and optical cameras plus temperature, radiation and oxygen sensors.** *(CERN/Alessandro Masi)*

RIGHT **TIM, parked up at its station in the CMS bypass tunnel.** *(CERN/Alessandro Masi)*

Chapter Four

Finding the Higgs and other discoveries

Ice thawed and snow had melted as the Large Hadron Collider came to the end of its winter shutdown period. It was early 2010, and it was nearly time to awaken the machine from its slumber. Whilst no proton beams raced through its tunnel, engineers had taken the opportunity to repair the particle smasher. After all, given the collider's magnet quench and other electrical faults that had damaged several of its superconducting magnets and contaminated its ring, the machine had seen better days.

OPPOSITE The first proton-lead collisions seen by ALICE in September 2012. *(CERN)*

ABOVE On 19 March 2010 the Large Hadron Collider's twin beams broke their energy record, reaching 3.48TeV per beam (~7TeV) during its first science run. That record has been broken multiple times since, with the current record standing at 13TeV. *(CERN/Maximilien Brice)*

CERN engineers already had a plan in place for blitzing the problem areas and, whilst it slept, there was also upgrading to do – the particle physics laboratory wanted their machine to be stronger and much more powerful. To get it there, higher electrical currents needed to course though the collider's magnets, and for that its entire system would be under greater demand. 'An intensive effort ensured that this work was undertaken and completed in the first three weeks of January, so that the hardware-commissioning teams could proceed with testing the magnets up to 6kA [which is over 11,000 times the current a household light bulb draws],' says James Gillies, head of CERN's communications. 'Several thousand channels of new quench-protection system were verified and measured precisely the resistance of 10,000 splices connecting the magnets. No unacceptably anomalous values were found.' The CMS experiment was even treated to an upgraded water-cooling system to ensure that it would run at optimum performance.

As nuts turned and further tests were run, the machine was ready to be powered up again. This time it would operate at 3.5TeV per beam, and it would be the accelerator's longest run, smashing particles right through to autumn 2011.

The early hours of 28 February saw the first beam circulate around the Large Hadron Collider since shutdown the previous year, metaphorically almost waving a flag to signal the beginning of an ultra-high energy race to delve deep into the realms of particle physics. CERN were getting ready for collisions that March, and at the end of that month scientists were crammed into the control room, anxiously waiting. Would the months of upgrading and maintenance pay off?

At first they didn't. Anxiety quickly turned to frustration as the day failed to go according to plan. Physicists had mapped out four points in the tunnel where they intended the counteracting beams to meet, putting on a show for the detectors and the teams that eagerly awaited them. The proton-packed beams didn't miss each other once, but twice, as they accelerated around the Large Hadron Collider at just a whisker short of the full speed of light. Looking deeply into the problem, the particle smasher's operations team found the source of the problem: on the first attempt, a power unit had tripped, and on the second, the

LEFT Physicists get a front-seat view of the lead ion run in the CERN control room. *(CERN/Maximilien Brice)*

accelerator seemed to overreact; the magnet protection system that had been installed to avoid another quench made the collider overly finicky about stray currents.

But their patience was eventually rewarded and, not long after midday, crossed arms and frowns turned to clapping hands and smiles as proton hit proton, creating the highest-energy collision that the world had ever seen. 'It was a great day to be a particle physicist,' says the laboratory's Director General, Rolf Heuer. '[People had] waited a long time for this moment, but their patience and dedication started to pay dividends.'

A new flavour of particle beam

Now that the collider was up and running, physicists were gearing up for their next big particle crash. But it wouldn't be protons in the collision; it would be lead ions – lead atoms stripped of their negative electrons. It took the collider's operations team a mere four days to make the switch, after the protons ran dry by November. For now, physicists were closing the door on simple proton physics.

As you can imagine, crashing lead ions together is quite different to making humble protons collide head-to-head. For one thing, the Large Hadron Collider has to be trained in order to handle another flavour of beam inside its chambers, by first threading one around

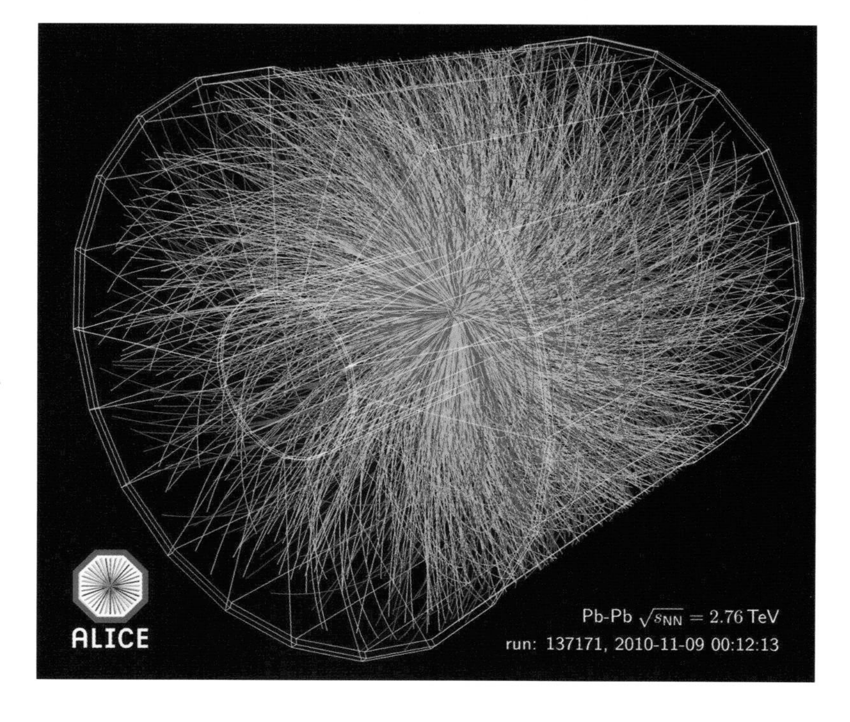

BELOW A collision between lead ions, as detected by ALICE in November 2010. The collision produced a whole mass of secondary particles: 1,209 positively charged particles (the darker tracks) and 1,197 negatively charged particles (lighter tracks). Eight per cent of the particles are pions. *(CERN)*

the clockwise direction before lacing a beam in the anticlockwise direction. The beauty of upgrading to lead ions is that you get more bang for your buck in terms of protons; each lead ion contains 82 of these particles, allowing for a highly intense, full energy of 287TeV almost effortlessly. 'Lead-ion running opened up an entirely new avenue of exploration, probing matter as it would have been in the first instants of the universe's existence,' says Gillies. '[Its aim] was to produce tiny quantities of such matter, which is known as the quark-gluon plasma, and to study its evolution into the kind of matter that makes up the universe today. The exploration [attempts] to shed further light on the properties of the strong interaction, which binds quarks into protons and neutrons.'

Most of the detectors on the Large Hadron Collider would sit out the show for this side of the experiment, but not ALICE, ATLAS and CMS – they couldn't wait for the lead-smashing show to begin. There was just another winter shutdown and proton-beam run to get past first.

The Large Hadron Collider makes a primordial soup

'There's a great deal of excitement at CERN today,' said Sergio Bertolucci, CERN's Director for Research and Scientific Computing, 'and a tangible feeling that we're on the threshold of new discovery.'

Bertolucci's announcement came just after midnight on an evening in April; the Large Hadron Collider had set a brand new record. Its beam intensity had thrashed that of the US Fermi National Accelerator Laboratory's Tevatron collider, a circular accelerator just east of Batavia in Illinois that was powered down for the last time in 2011. 'Beam intensity is key to the success of the Large Hadron Collider, so this was a very important step,' explains Heuer. Luminosity is useful when working out how many collisions have just happened inside an accelerator; the more vibrant the luminosity, the higher the number of particles that have crashed into each other – it's a very important piece of information to know when seeking out the rare birth of new kinds of particles. 'Higher intensity means more data, and more data means greater discovery potential,' adds Heuer.

And the conditions to make those discoveries came thick and fast. 'Besides black holes, there's nothing denser than what we've created,' said physicist David Evans, a team leader for the ALICE experiment. His 10,000-tonne detector had generated temperatures more than 100,000 times hotter than our Sun; lead ions – carrying protons and neutrons – had melted, freeing the quarks from their bonds with gluons. The Large Hadron Collider had recreated conditions similar to those after the Big Bang. It had made the quark-gluon soup that was thought to have filled a young universe

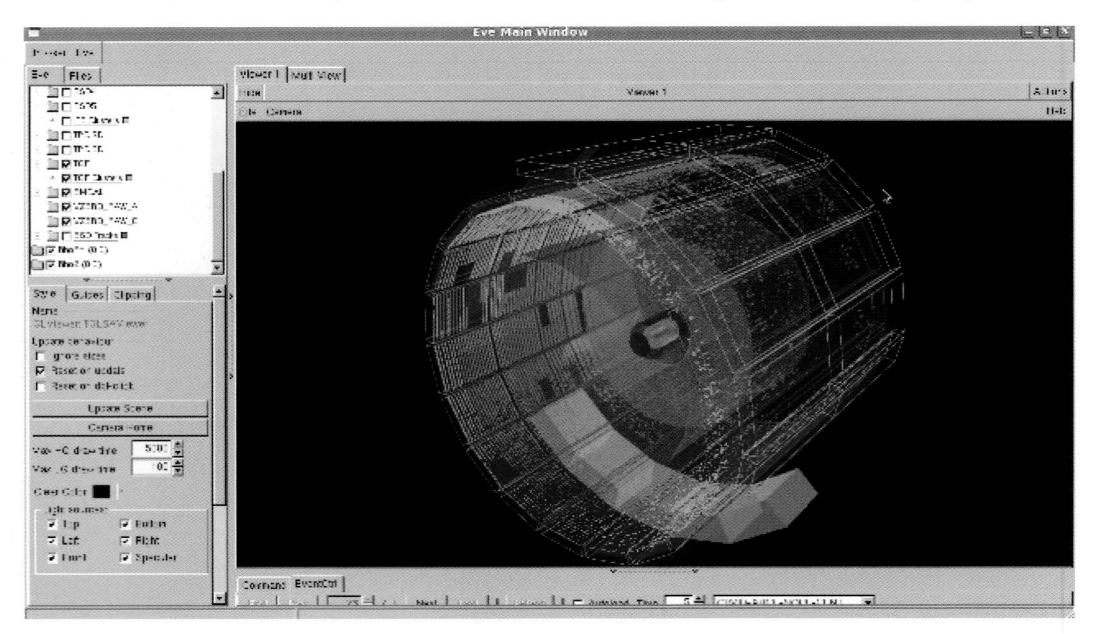

RIGHT The ALICE experiment sees its first ions during a lead beam injection test. *(CERN)*

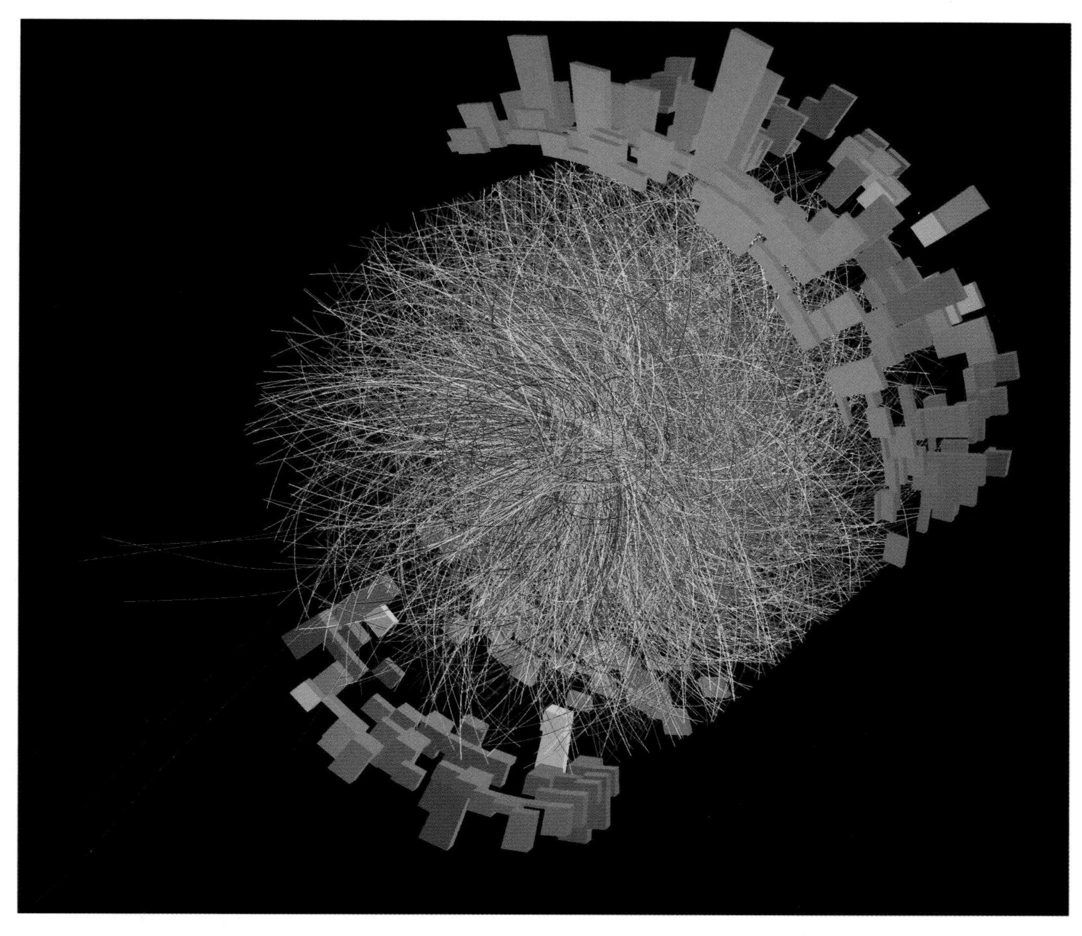

LEFT Heavy ion events seen by ALICE in November 2015, which produce quark-gluon plasmas. *(CERN)*

that was only a trillionth of a second old. During this time the quarks and the gluons would have been widely spaced before the soup cooled, connecting up to form the matter we see today.

There's another speculation about this plasma too, and that's that it acts more like a gas – rather than the liquid you're most likely picturing at this moment – with quarks and gluons spaced out much more. ALICE turned up evidence that could point to a more 'airy-like' behaviour, given slight differences between the particle collider's results and those revealed by the Relativistic Heavy Ion Collider (RHIC), whose home is at the Brookhaven National Laboratory, New York. 'It could well be that in the very early stages [of our quark-gluon plasma], it's behaving more like a gas, and then as it cools it turns into a liquid,' explains Evans. 'We still need to investigate this further.'

Nevertheless, RHIC – which turned in its results in 2010 after crashing gold ion beams together, creating a temperature of four trillion degrees Celsius (seven trillion degrees Fahrenheit) – had given physicists a lead, a snapshot of how the quark-gluon plasma evolved as the universe aged. 'I think we're now at a point where, with these two machines,

BELOW The Relativistic Heavy Ion Collider (RHIC) at Brookhaven National Laboratory (BNL) in the United States. *(Brookhaven National Laboratory)*

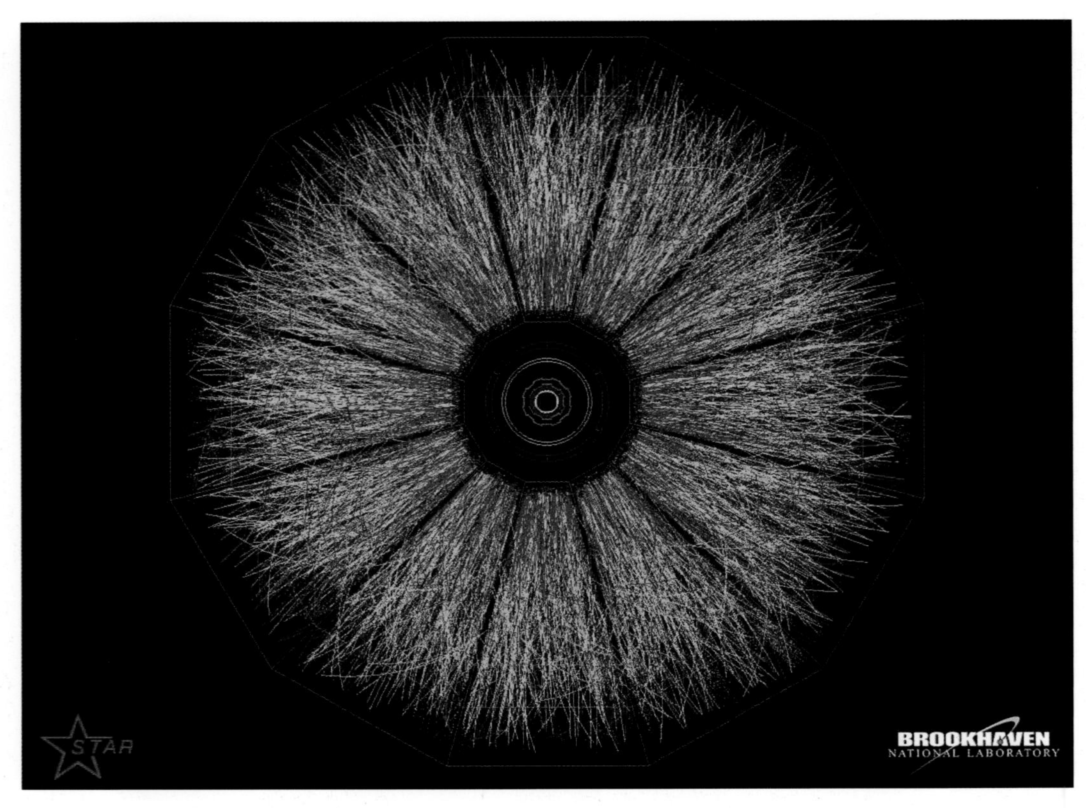

RIGHT Results from the Relativistic Heavy Ion Collider when gold ions were collided. *(CERN)*

we can look over a wide energy range at the properties of the quark-gluon as it evolves with temperature and density,' Thomas Ludlam, chair of the physics department at Brookhaven said at the time, poring over the results of both particle accelerators. 'Now that we'd all shown our hands,' said ATLAS' Peter Steinberg, 'we're going to start comparing results.'

In looking carefully at how particles respond to the plasma, Steinberg and others had been given tantalising information about the quark-gluon mixture, more specifically hints at its temperature and how much influence it truly had in the early universe. The particle that made it all possible was the upsilon, a meson made up of a quark and an antiquark. These particles are often found in three bound states, that is they're pretty much identical save for the strength of the bond between the quarks that make them.

In May 2011 CERN had made a breakthrough of sorts. They'd discovered that some upsilons seemed to melt apart in the plasma, and knowing the energies at which this happened had given them an insight into working out how hot it was – which showed that the quark-gluon soup was much hotter than the weakly bound upsilon, but didn't have the ultrahot-oomph needed to divide the quarks inside the stronger upsilon. Meanwhile, teams behind ALICE, ATLAS and CMS were digging around in the data; they were interested in how other particles felt about being around primordial soup.

It turns out that the J/psi meson – made up of a quark and an antiquark in the charm family – fell apart in the quark-gluon plasma. Meanwhile, the packets of light known as photons and Z bosons didn't seem at all fussed about being stuck in the Large Hadron Collider's creation.

BELOW Phase diagram of the quark gluon plasma. *(CERN)*

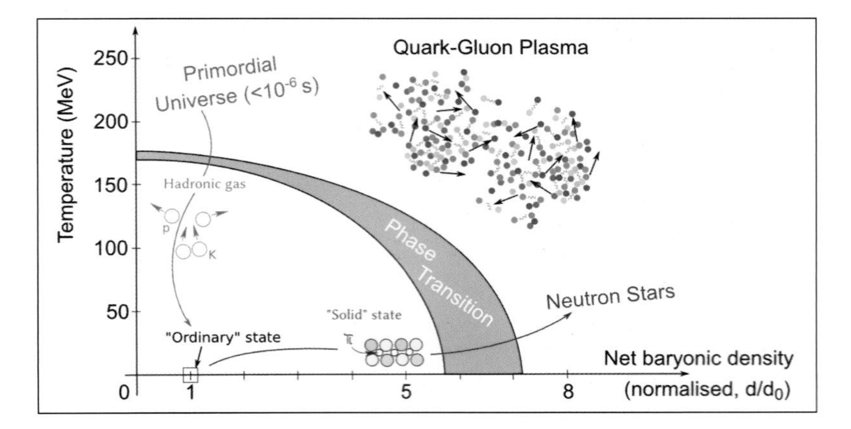

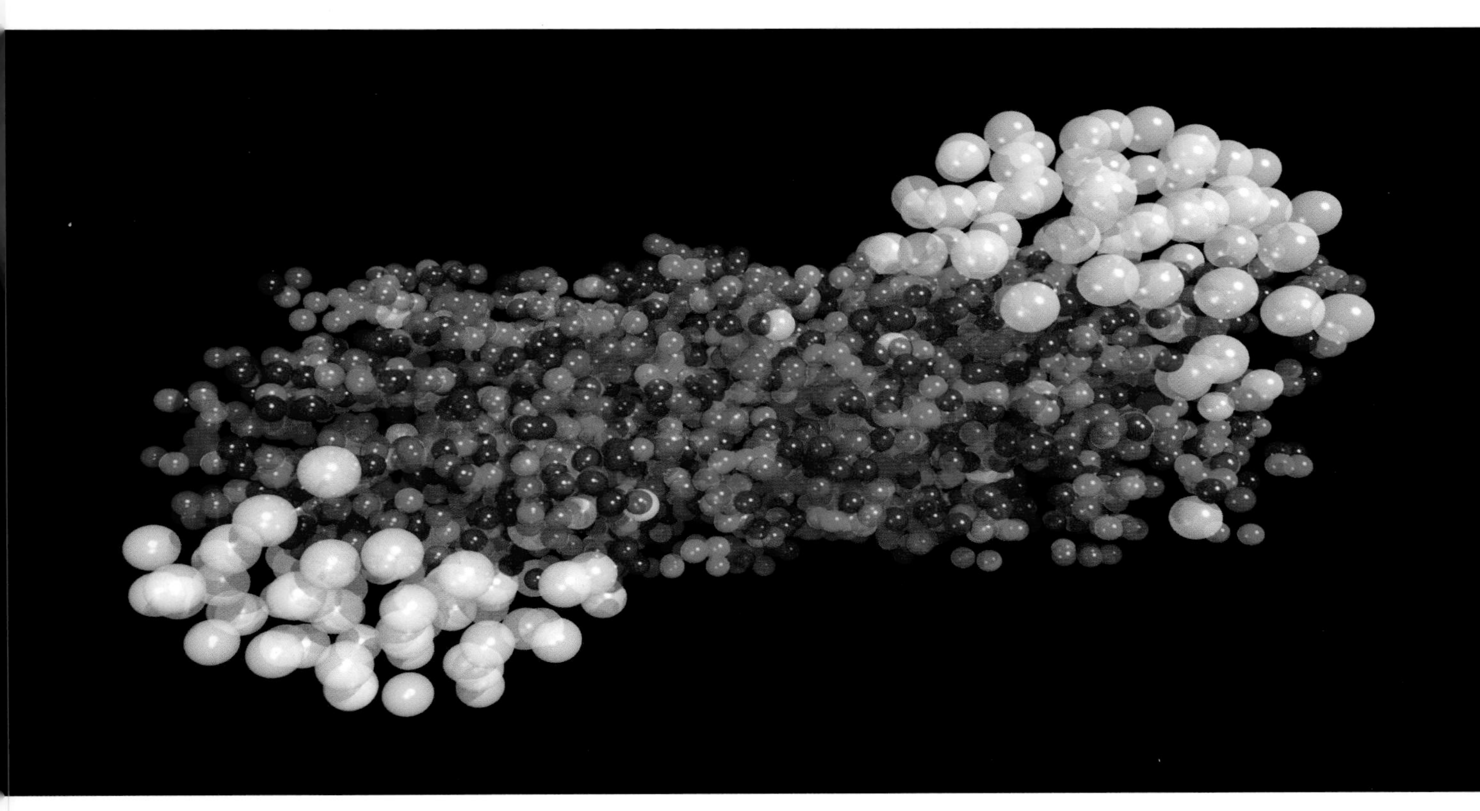

ABOVE A representation of a quark-gluon plasma, such as that formed in experiments at the LHC. *(CERN/Henning Weber)*

Whilst lead ion beam crashed into lead ion beam, physicists watched intently. They figured that they could learn a lot from the jets, produced by a process known as jet quenching, and their computer screens revealed what they wanted to know; back-to-back jets made inside the plasma raced in all directions. Something interesting they noticed immediately was that the jets didn't have a great deal of balance. Where the plasma was thick, packed full of particles, the jets seemed to shrink away before making a break for it on the other side. It was like they were being momentarily zapped of their energy, completely consumed by teeming quarks and gluons. 'The Large Hadron Collider experiments had taken a giant step toward understanding the properties of the [plasma],' says the ALICE experiment's John Harris. 'To me that just jumped out. We'd made a quantum leap.'

The hunt for the Higgs

CERN had the machine. As far as physicists were concerned, it was open season for hunting the Higgs; the missing piece of the Standard Model that's responsible for all of the mass in the universe. If they found it, it wouldn't just be one of the greatest discoveries in the past 50 years, it would explain once and for all why each and every particle that makes up everything you see around you has the mass that it does.

Underground at the Large Hadron Collider, particle beams continued to crash and more and more data continued to spill outward. Eager physicists carefully and painstakingly sifted through this data, searching for a rare process that could be the smoking gun. They'd only collected a mere tenth of the total data, but slowly and surely – and combined with the efforts of previous colliders like the Large Electron-Positron collider and the Tevatron – physicists were leaving the Higgs with nowhere to hide. 'While it was too early for the biggest discoveries, the experiments had already accumulated interesting results,' said Heuer. But CERN did have something to show the world.

Heuer would make the announcement at a conference in Grenoble, France, alongside scientists behind the New York collider. They'd ruled out several mass ranges, backing the Higgs into a corner. The more beams they smashed together, the more they were sure that the so-called 'God Particle' couldn't exist at masses between 157 and 174GeV. The LEP had already pipped its successors to the post in ruling out the possibility of an exceedingly light Higgs. Physicists crossed out a mass of 115GeV.

The Large Hadron Collider might have collected less data than Fermilab's Tevatron collider, but it was head and shoulders above it in terms of power. It would be ATLAS and CMS who would whittle down the field even further. 'The Large Hadron Collider had finally entered the Higgs game, giving the first direct constraints in the high-mass region ever reached by experiment,' the ATLAS experiment's Dave Charlton told a conference-room packed full of particle physicists. 'The data was only collected up to [about] three weeks ago so there's much we still have to look into; if the Higgs exists, then we've now got hints that we should be looking more carefully at the 130 to 150GeV region.'

It was a matter of flicking through several PowerPoint slides to narrow down the search. But there was something else – both CMS and ATLAS had spotted what CERN had dubbed 'excess events', when sniffing around low mass ranges in the hunt for the Higgs. The announcement had piqued the interest of the room's occupants; it could have been a fluctuation in the background, the way that the Large Hadron Collider's detector duo were looking at the background noise behind the collisions, or something much more exciting. 'It was also precisely what you would expect to see if a low-mass Higgs was starting to show itself,' says the physics coordinator of CMS, Gigi Rolandi. What's more, the data bursts weren't visible at higher masses by either Rolandi's experiment or ATLAS.

There was no doubt about it, CERN were truly in discovery mode; but some physicists weren't ready to get excited just yet. 'It was too early to get carried away,' Rutgers University theorist Matt Strassler quickly pointed out. 'It was vital that the background, especially events where two W bosons are produced, was correctly understood.' As you'll recall, the Higgs likes to break down into W bosons, which leave behind a signature that can be picked up by the Large Hadron Collider. Confusingly, W bosons can also be made when quarks decay. As you can imagine, telling the two processes apart is very difficult. If they got it wrong, then it would seem like there was an enormous burst of Higgs-like events; something that was near-impossible.

While not being able to move data – with the Large Hadron Collider's Computing Grid, routinely processing up to 200,000 analysis jobs concurrently for computer centres all over the world – a flurry of plots revealed that what we had always known about the Standard Model was true. Physicists could feel that they were getting closer to finding out about the myriad exotic black holes and states that are thought to spring from the extra dimensions slicing up the universe. Even bosons that carry forces as they travel, amongst other hefty particles, were running out of places to hide, giving the physicists a window beyond the Standard Model. Even the unassuming quark delivered as expected, appearing as a single point in the data at a range of energies. However, they hadn't spotted supersymmetric particles just yet, which you'll remember to be the 'equal-but-opposite' copies of particles slotted into the Standard Model.

But, regardless, we were closing in on a region of particle physics like never before. There was one common statement amongst the particle physicists in attendance at the Europhysics Conference of High-Energy Physics as they filed in and out of sessions: it was getting real. 'Assuming the Large Hadron Collider keeps running as it is, we will know by the end of October whether the Higgs exists or not,' grinned Rolandi. 'Within the next year we will have, one way or another, a completely different view of Nature which will have major consequences for the [particle physics] field.'

Meanwhile, at Point 8, LHCb was tackling B-mesons in order to figure out what caused the imbalances between matter and antimatter at the dawn of time; in particular the disintegration of these particles and their antimatter partner, the B-bar. Common in the commotion of the Big Bang, these two are made up of quarks, more specifically the bottom quark and an antiquark. LHCb found the pair to decay differently. This was the behaviour CERN needed to get a sliver of information about the symmetries found in Nature and the fundamental forces that exist within it – as it stood, the particle pair were

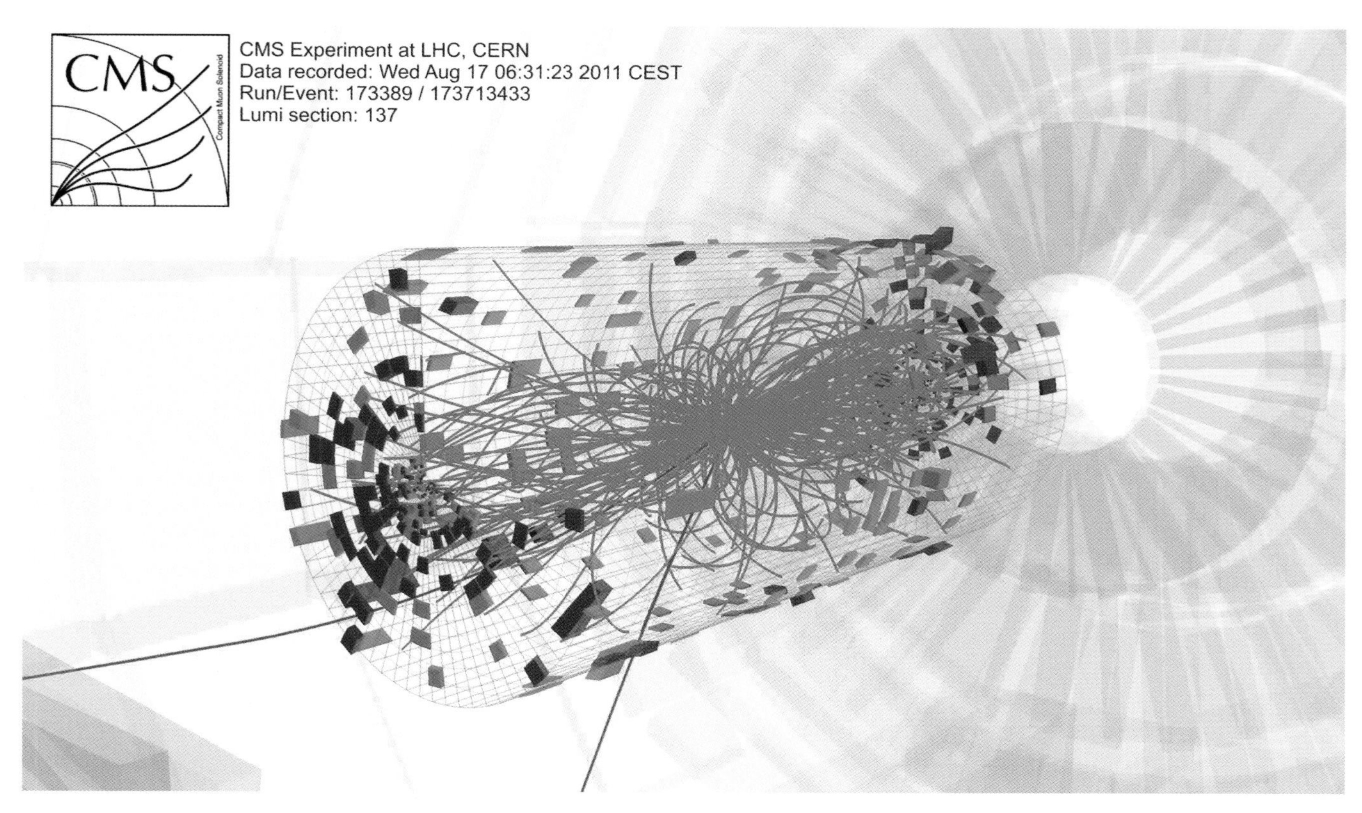

ABOVE Di-muons, represented by the red line extending from the particle collision, as seen by the CMS in 2011 as it probed the secrets of matter and antimatter. *(CERN)*

violating something called charge conjugation parity or CP symmetry. In a nutshell, this rule states that even though matter and antimatter are opposites, they should also exist in equal measure. Look around the universe: you, this planet, the stars and galaxies, all suggest that something went awry somewhere; CP symmetry was violated.

'We wanted to figure out the nature of the forces that influence the decay of these [B mesons],' said Syracuse University's Sheldon Stone, who is on the LHCb team that made the exotic particle in millions of proton-proton crashes. 'These forces exist, but we just don't know what they are. It could help explain why antimatter decays differently to matter.'

Months later, particle physicists collaborating on the experiments at the Large Hadron Collider gathered in Mumbai for the Lepton-Photon conference. Word on the particle physics circuit was that the Higgs hadn't been found. In fact ATLAS and CMS had excluded its existence over a wide mass range, and with a 95% certainty. But physicists weren't worried; the way they saw it, every particle that crashed with another under the border of France and Switzerland was bringing them closer to where the God Particle was hiding. After all, they knew where not to look – and that was over the mass region of 145 to 466GeV. Even if they found that the boson didn't exist, then that would be another cause to pop the champagne corks. 'These were exciting times for particle physics,' recalled CERN's research director, Sergio Bertolucci. '[At this time] if the Higgs existed, [we felt that] the Large Hadron Collider experiments [would soon] find it. If not, its absence would have pointed the way to new physics.'

New physics or not, the fluctuations that had been announced in France were neither that nor the elusive subatomic particle. They'd gone, taking any tantalising hints and ripples of excitement with them. As proton–proton and lead ion–lead ion beams continued to collide, CERN had gathered twice as much data, fading the 'excess events' into obscurity. Originally the ATLAS and CMS experiments had hinted that the Higgs was somewhere around the mass range between 120 and 140GeV, but now it seemed like the particle didn't want to be found. 'We were close to a depressing moment in which we concluded that those fluctuations were statistical jokes, but there was also the possibility of seeing them grow with more data,' remembers CMS spokesman Guido Tonelli. It was true – in particle colliders it's common for blips and fluctuations in data to suddenly arrive and disappear again.

'We had to be patient,' said head of the ATLAS detector group, Fabiola Gianotti. 'We needed to take the data, analyse them and understand them. At the same time, we were super excited, because we were very close. We were months away from really solving one of the major mysteries in fundamental physics. It was so close I felt I could touch it with my hand.'

Game over for supersymmetry?

Things were not faring much better for LHCb, which was grappling with another conundrum: it couldn't find any evidence for the supersymmetry that's thought to exist throughout the universe. As far as the experiment was concerned, no super particles existed, no partners for each and every member of the Standard Model. No link between the two different fermions and bosons, no explanation for dark matter, and no assisting the Higgs in explaining why particles have mass. The decay of B-mesons had told us nothing.

Was it game over for the theory? 'It did rather put supersymmetry on the spot,' confessed Tara Shears, spokesperson for the LHCb experiment. After all, if these particles did exist, the B-mesons would have decayed much more often than what the Large Hadron Collider detector had revealed. Across the pond, the Tevatron accelerator had hinted that B-mesons fell apart, since supersymmetric particles were knocking around.

It just didn't seem to make any sense; had a theory – at the time developed some 20 years ago – finally bitten the dust? 'There was a certain amount of worry that was creeping into our discussions,' Fermilab's Joseph Lykken revealed. 'It's a beautiful idea. It explains dark matter, it explains the Higgs boson, it explains some aspects of cosmology. It didn't mean it was right, though.'

But CERN and its particle physics collaborators weren't prepared to close the door on supersymmetry for good. To them, it was too beautiful a model to let go completely, plus the theory needed to be tested within an inch of its life before textbooks had to be rewritten. It felt like there were still things the Large Hadron Collider could try.

First particle discovery

November saw the Large Hadron Collider juggling magnetic fields that would collide protons and lead for the very first time. Previously, the particle physics laboratory hadn't been mixing the beam flavours, but scientists were keen to know as much as possible about the lead ions before they were obliterated in a head-on collision – something they could only figure out by going in cold, and before the two beams met. The proton would act as a probe; years of experiments and try-outs had told us enough about the quarks and gluons inside a proton, so anything odd that came about during this hybrid collision could be blamed on the lead ion.

CERN were keen to find out more about the so-called parton distribution function, something they couldn't delve into using lead ions alone. If you're not familiar with what a parton is, it's the quarks and gluons that make up hadrons – also known as more familiar protons and neutrons – and which compose the lead ions. You can imagine these hadrons to be quarks linked by gluons at low energy. Inside the Large Hadron Collider, though, energies are so intense that hadrons can actually be made up of extra partons that have a say in how the collisions play out.

The very first beam-to-beam smash lasted for 16 hours, but it was worth the wait; lead bunches were injected into the collider's ring as 304 proton packets raced in the opposite direction. The energy was ramped up, smashing a few bunches of each particle type around the accelerator at a very relativistic 3.5TeV for protons and 287TeV for lead, or more specifically 1.38TeV for every lead ion. The two hadn't collided, but the test was a success, allowing scientists to work out when and where they wanted the beams to clash.

There would be more tests ready for the duo of beams the following year. Not convinced that they'd left every stone unturned, CERN went back to crashing lead ions head to head.

Just gone midnight – by three minutes, to be exact – after troubleshooting at the Large Hadron Collider's controls to ensure that the beams were in tip-top condition and ready for physics, radiation known as synchrotron light was picked up by the telescope and accelerator system packed inside the accelerator. It told the physicists that the lead beams had picked up twice the energy they were injected with on their race around the accelerator. 'We see a shimmering spot on the screen,' says physicist John Jowett. 'It's as close as we get to seeing the beams of lead nuclei with our own eyes.'

There was a buzz about the particle physics community by the close of December. ATLAS had found something – a peak in a spectrum plot turned out by the experiment's detectors; it was the evidence for a brand new particle that had never been seen before. And it wasn't just any old particle; it would help scientists to have a better understanding of the strong forces that hold matter tightly together. 'This was the first time such a new particle had been found at the Large Hadron Collider,' ATLAS' Paul Newman explained. 'Its discovery was a testament to the very successful running of the collider in 2011.'

Newman is talking about the Chi-b (3P) particle, which belongs to the boson family, the same family as the sought-after Higgs. Chi-b (3P) is made up of a beauty quark and its antiquark equivalent, linked by the strong force that's familiar in holding nuclei of atoms together. It was the perfect candidate in working out how even the smallest components of matter are linked together, building up everything we could ever see or imagine in the universe. 'The better we understand the strong force, the more we understand a large part of the data that we see, which is quite often the background to the more exciting things we are looking for, [which at the time was] like the Higgs,' says Roger Jones, who works on the ATLAS detector.

It's true that Chi-b (3P) had already been predicted, but it wasn't exactly how physicists had expected. Giving away its existence with the help of photons that transformed into an electron-positron pair as they crashed into material inside the detector, Newman and his team realised that the particle was heavier than expected, tipping the scales – as far as particle masses go – at over 20,000 times more massive than the electron. What this told the ATLAS team was that the quark-antiquark pairing was much more loosely bound by the force that exists between them.

There's a name for this bound state. It's called quarkonium. It isn't too different from what you expect in a stable hydrogen atom, with its nucleus made up of a positive proton and a neutral neutron, with an electron whizzing in orbit around the tightly-locked pair. The ATLAS team were waiting on the behaviour of photons in order to make their discovery – the same particle that's emitted when an electron inside a hydrogen atom drops through different energy states.

It was a proud moment for the network of scientists and engineers behind the most-powerful particle accelerator in the world. It would be later, making the finding all the more sweet, when the United States' Tevatron collider confirmed the Chi-b (3P) particle. The Large Hadron Collider had made its very first discovery; the world was just waiting for it to make its next.

The collider sleeps on the particle physics problem

Returning the Large Hadron Collider to winter hibernation mode allowed physicists a chance to ponder on how best to proceed with hunting down the Higgs. They were to-ing and fro-ing about which energy level to crank it up to. They wanted to find the particle, but didn't want the machine to be put at any risk.

Yet, the way they saw it, they were getting closer to finding its hiding place. Besides, the increase in energy was modest, but could provide a much greater chance of finding the particles predicted by supersymmetry, and even finding the Higgs boson once and for all.

It was a tantalising thought, and there was nothing else for it – it was time to kick up the beam energy once again, to 4TeV per burst, combining forces to 8TeV overall. 'The experience of two good years of running at 3.5TeV per beam gave us the confidence to increase the energy without any significant risk to the machine,' explains the Director for Accelerators and Technology, Steve Myers.

'Now it was over to the experiments to make the best of the increased discovery potential we're delivering them.'

The engineers were all too aware of how finicky the superconducting magnets that form the bulk of the collider were, given their experiences of the quench that had knocked the entire experiment offline. This time, though, they weren't taking any chances as they got to work remaking connections and upgrading the machine, making full use of the months mapped out before them. They had to ensure each and every part of the great machine was up to the job, ready to handle the all-new level of intensity.

While the particle accelerator slept, it was an especially intense period for the ALICE experiment and its team of physicists. Not only did the detector need to be maintained in general, but its hardware needed to be upgraded. At that point in time it wasn't up to the task that would be imposed by the Large Hadron Collider when it awoke.

In a nutshell, the ALICE team wanted their experiment to be able to 'see' more. Installation of several supermodules in the experiment's Transition Radiation Detector (TRD) and the Electromagnetic Calorimeter (EMCal) were the ticket; both would be able to snatch more coverage of collisions during the intensive run planned for the following year. 'We [wanted] to extend ALICE's physics reach, particularly towards investigating the thermodynamic properties of the quark-gluon plasma – the state of matter that is formed at the collision energy of the Large Hadron Collider,' explained Yves Schutz, a spokesperson for ALICE. 'We'd already observed intriguing results concerning the formation of particles issued from the plasma. These results needed further analysis.'

But before that could even happen, the ALICE teams were faced with technical issues. Filters packed inside the cooling system of the Silicon Pixel Detector (SPD) were on the blink; discharges of high voltage blew up in the Time Projection Chamber (TPC), and faulty low-voltage connections in the Muon Tracking Chambers were standing between CERN and switching the collider back on.

Engineers worked around the clock to get ALICE and the remainder of the machine up to speed before circulating beams around the tunnel once more. They didn't know it yet, but one of the greatest scientific discoveries of all time was just around the corner. A very special particle was about to put in an appearance.

BELOW **Dr Joe Incandela, scientist and spokesperson for the CMS experiment, whose team spotted the Higgs boson.** *(CERN/Maximilien Brice)*

A hint of the Higgs?

With his arms crossed, frowning at the data that was being presented to him, former spokesperson for the CMS experiment Joe Incandela wasn't convinced. His colleagues were trying to say that a bump in their data was the elusive particle that would stop the Standard Model from falling apart. Physicists had already uncovered the Chi-b (3P), but they couldn't explain the spike in their plot that the Large Hadron Collider experiment had created. They were explaining the anomaly as being the Higgs boson.

At the time, Incandela cast them aside as preliminary results; the collider was in the early stages of discovery, and these results from the CMS were likely to be nothing more than physicists testing the particle physics waters. After all, the Higgs could reveal itself in many different forms. Passing their find off as the holy grail of physics was foolish. 'A tiny mistake or an unfortunate distribution of background events could make it look like a new particle is emerging from the data when, in reality, it's nothing,' Incandela explains. 'We'll never be

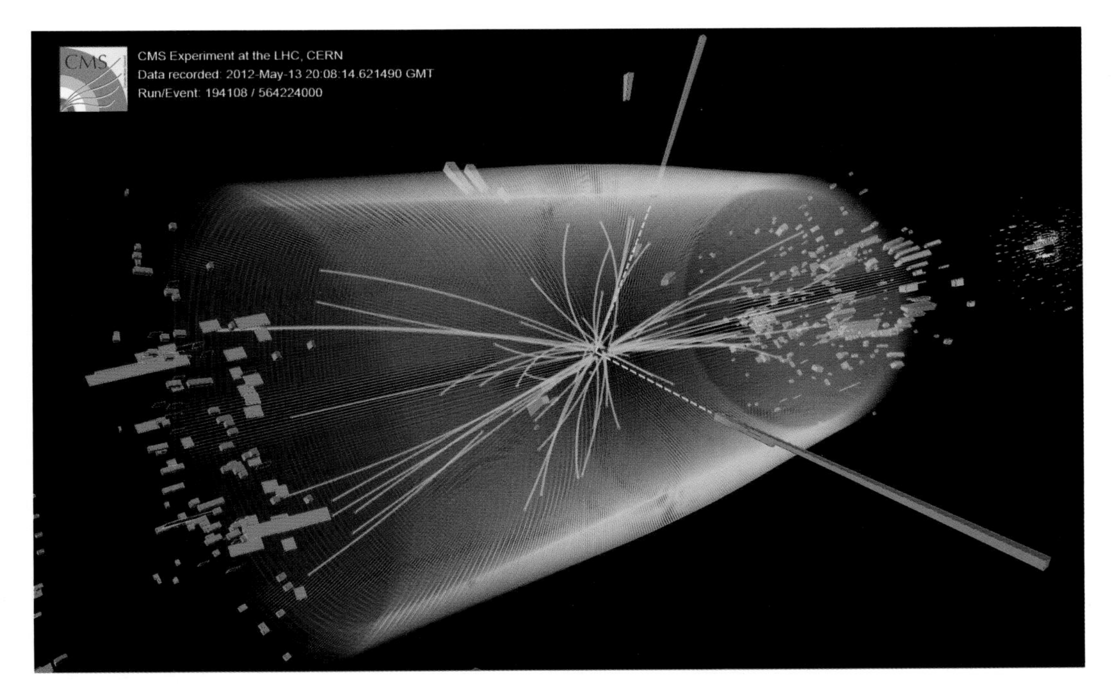

LEFT A collision between two protons, with a combined mass-energy of 8TeV, seen by the CMS. Two photons, indicated by the yellow and green dashed lines, are thought to have been from the decay of the Higgs boson. *(CERN)*

able to definitively say if something is exactly what we think it is, because there's always something we don't know and cannot test or measure.'

Incandela continued to dig in his heels. He wanted a retrial on the results. Each and every member of the team would refine their data analysis, complete with fake samples of data spat out by computer simulations that would cover up the more interesting segments of their studies. When they'd worked through their method of translating the information, along with other data to make a decent observation, they could then rip off the patch of false data and be left with an algorithm they could trust. 'It was a nice way of providing an unbiased view of the data and helps us build confidence in any unexpected signals that may be appearing, particularly if the same unexpected signal is seen in different types of analyses,' remembers Incandela.

But the bump that he saw all those months ago wasn't going away. In fact, it was even more significant, proudly sticking out of the data plot at the same mass range, even when Incandela fed the data through two separate channels. 'I knew we had something,' he said. 'We presented the results to the rest of the collaboration. The next few weeks were among the most intense I have ever experienced.'

It wasn't just the CMS that had sniffed out hints of the Higgs. The team behind ATLAS were busy studying their data too. Andrew Hard, who was responsible for writing code for the experiment that gives it the power to pick out and calibrate the particles of light that the detector records during high-energy collisions, skipped his Christmas holiday to wrestle with the new data. 'There were a few days when I didn't see anyone else at CERN,' Hard said. 'I thought some colleagues had come into the office, but it turned out to be two stray cats fighting in the corridor!'

BELOW CMS' discovery of the Higgs boson. The little bump in the graph above the background signal is the signature of the Higgs. *(CERN)*

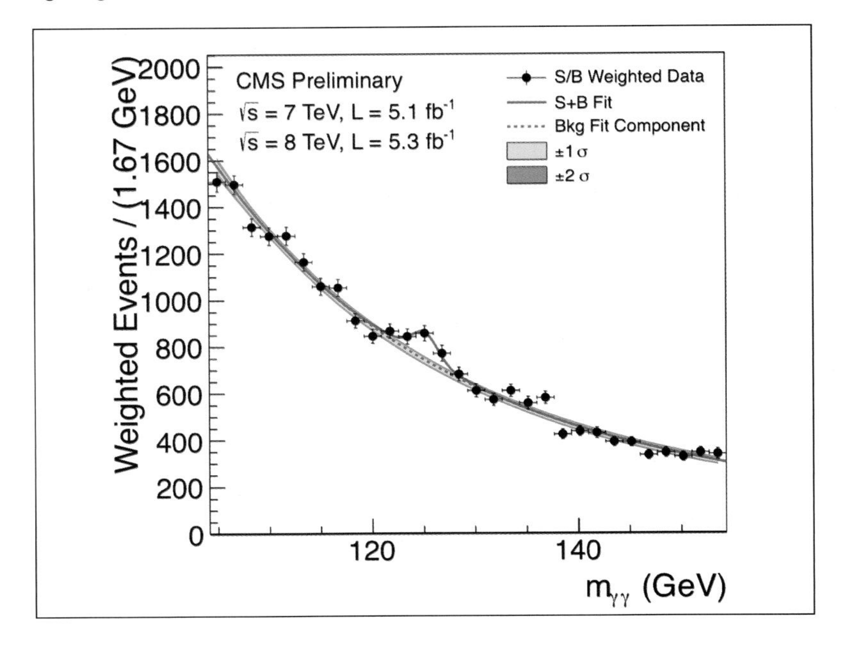

ABOVE Dr Fabiola Gianotti, scientist and spokesperson for the ATLAS experiment. *(CERN/Christian Beutler)*

Hard knew that the Higgs could transform into two particles of light when it decays, so he was especially interested in the data bump. The same spike that the CMS team attempted to wash out of their data also wouldn't budge from that picked up by ATLAS. 'People collaborated well and everyone was excited about what would come next,' he said. 'All in all, it was the most exciting time in my career. Everyone in the group started running into the office to see the [bump] for the first time. They took a bunch of photos [of the plot].'

CERN makes the discovery of a decade

'When this big excess showed up in July 2012, we were all convinced that it was the guy responsible for curing the ails of the Standard Model, but not necessarily precisely that guy predicted by the Standard Model,' said Matthew McCullough of the Massachusetts Institute of Technology, who was watching the drama unfold.

Yet with the excitement came a whole new world of stress, nervousness and anxiety for physicists behind CMS and ATLAS. 'One of our meetings [where we scrutinised the data] lasted over ten hours, not including the dinner break halfway through,' said Hard. 'I remember getting in a heated exchange with a colleague who accused me of having a bug in my code.'

But after checking, double-checking and cross-checking, both Incandela and ATLAS' spokesperson Fabiola Gianotti realised that they had something special; they were so confident

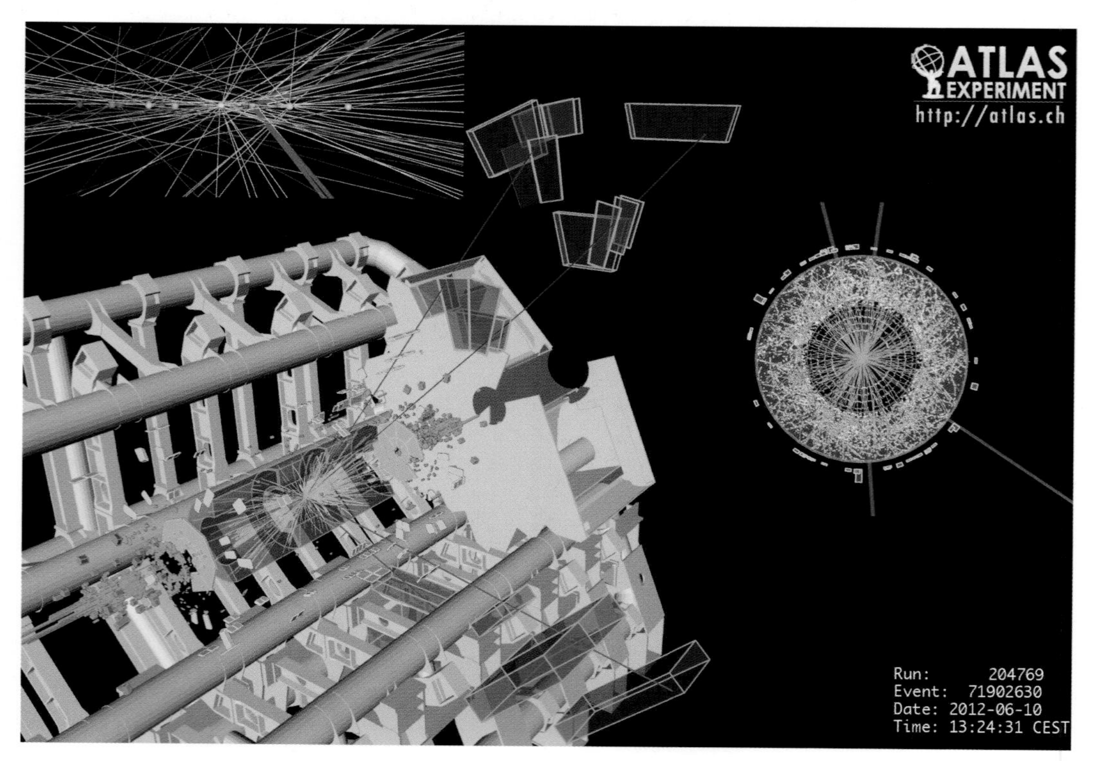

RIGHT ATLAS' view of a Higgs boson decaying into four muons. *(CERN)*

that they slapped a five-sigma on the discovery. Five-sigma is the strongest level of confidence scientists have in their results; what it says is that the probability of a statistical fluke being capable of creating a hump in the results that the CMS and ATLAS saw is actually less than one in three and a half million. Think of tossing a coin; a five-sigma would correspond to getting 21 heads in a row, while a three-sigma has the likelihood of tossing nine.

That was enough for Incandela and Gianotti. On 4 July they had gone ahead with their decision: they were ready to tell the world that they'd found the particle-decay trail of the Higgs boson, weighing in at 125.3GeV, about 133 times heavier than the proton. In a packed

ABOVE LEFT Peter Higgs visits the LHC, where the boson whose existence he had postulated was discovered. *(CERN/ Mona Schweizer)*

ABOVE The Director General of CERN, Rolf-Dieter Heuer, informs staff at the LHC of the Nobel Committee's award of the 2013 Prize in Physics to Peter Higgs and François Englert for the LHC's discovery of the Higgs boson. *(CERN/Maximilien Brice/ Anna Pantelia)*

LEFT The Belgian physicist François Englert, who co-developed the theory behind the Higgs boson along with Peter Higgs, at the LHC. *(CERN/ Claudia Marcelloni)*

ABOVE **It was standing room only for the 2012 press conference at CERN announcing the discovery of the Higgs boson.** *(CERN)*

CERN auditorium, swamped with scientists and the press, one man wiped tears of joy from his eyes: Peter Higgs, who – alongside six other physicists – had proposed its existence in 1964.

Another crack at supersymmetry

Behind the excitement of a major discovery, the LHCb stood in the shadows. Its team were waiting to make an announcement of sorts of their own. You'll remember that months before, supersymmetry had been momentarily cast aside as physicists worked their models and took more data from the increasingly energetic Large Hadron Collider? Well, now results from the LHCb were back with a vengeance, thrusting the experiment into the spotlight. What's more, they had provided a glimmer of hope that the theory of 'super particles' could actually work, even if just slightly – at least in the minds of some particle physicists.

RIGHT **Peter Higgs (left) and François Englert at the 2013 Nobel Prize press conference at the Royal Swedish Academy of Sciences.** *(CERN)*

ABOVE The 'Long Shutdown', as the LHC's two-year hiatus was known, allowed for numerous upgrades. Here, the beam pipe is removed from the CMS for refurbishment. *(CERN/Michael Hoch)*

'Supersymmetry may not be dead but these results had certainly put it into hospital,' said Chris Parkes, a spokesperson for the experiment. While many had hoped that the particle accelerator would have confirmed the theory at this point, there was every possibility that the team had chanced upon new physics. Other researchers believed that the LHCb's latest presentation was what you would expect when hunting down evidence for supersymmetry. 'I certainly wasn't going to lose any sleep over it,' physicist John Ellis of King's College London said in the theory's defence.

LHCb had been watching the decay of B-mesons into smaller muons, in what had been dubbed the ultimate golden test for supersymmetry. On the face of it, it felt that the theory had failed the exam, but there was no denying that watching the mesons fall apart into previously unseen particle-antiparticle constituents wasn't unexciting in itself.

The LHCb team had done their sums. They'd worked out that for every billion times the meson decays, it only becomes a muon and an antiparticle three times, giving it a statistical level of 3.5-sigma. It wasn't 'discovery-worthy', but it had undermined an important theory, and that in itself was worthy of further study. But the experiment had run out of time on this run. The Large Hadron Collider was getting ready for its big shutdown – this time for a lengthy two years.

When protons meet lead ions

Before the particle accelerator powered down, ceasing the whizzing of beams in its 27km tunnel, there was one final show it wanted to put on for its audience of detectors: the high-energy jousting of lead ions and protons. CERN had caught a glimpse of the two species making their way around the particle smasher during test runs. 'Compared to the lead-lead collisions, you're looking at different physics at different beam intensities,' explains Christof Roland, who was the co-convener of the CMS' Heavy Ions (HI) group. 'You need dedicated trigger menus. This was where the biggest effort went in from the HI group: to prepare studies of all the triggers for the proton-lead run.'

When the physicists crashed lead-nuclei together, there were 4,000 interactions for every second. The CMS experiment captured about 200 of these to tape. Meanwhile, the proton-lead collisions are much more volatile, smashing together two million times every

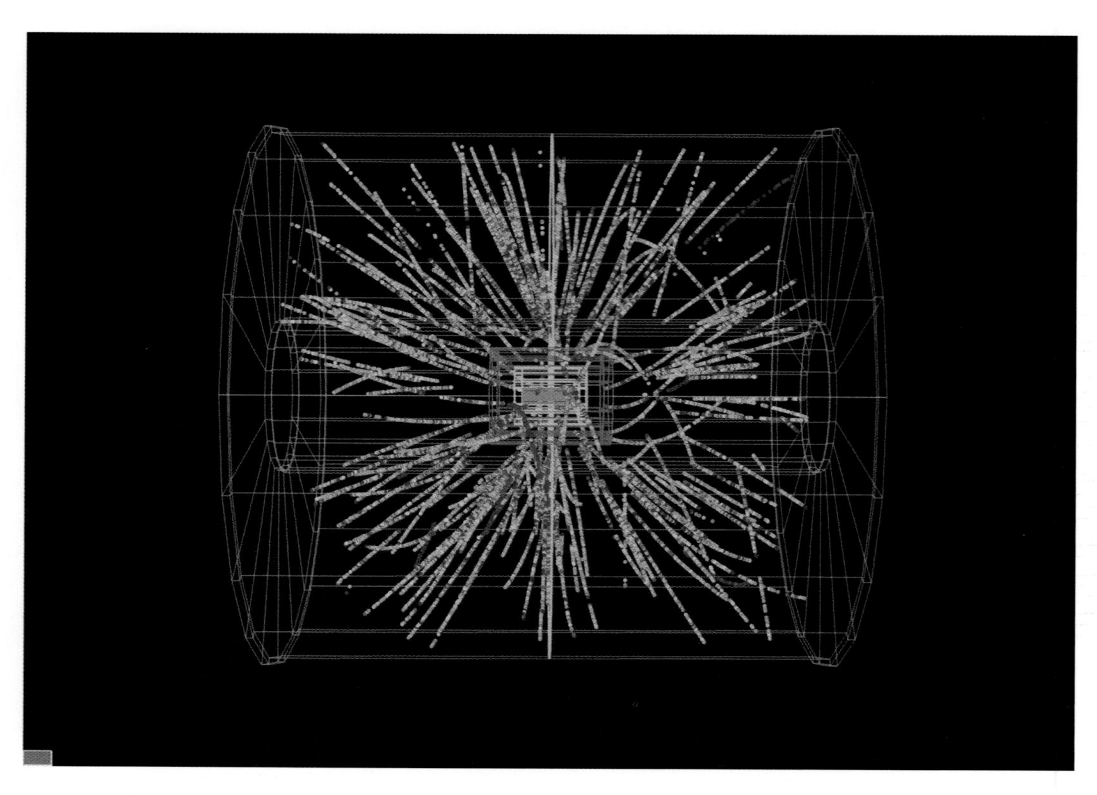

RIGHT An event display during collisions between lead ions and protons generated by ALICE's High-Level Trigger (HLT). *(CERN)*

second, allowing the CMS to capture about 1,000 subatomic fireballs. 'The pilot run was extremely useful when getting an idea of what the rates would be so we could optimise everything,' says Christof. Of course, the collision rate between protons is much higher inside the Large Hadron Collider. Smash-ups between protons are actually fairly harsh, so the CMS experiment had felt the brunt of previous particle physics activity. It was time for an upgrade of its CASTOR and ZDC subdetectors; they would ensure that the experiment wouldn't miss a trick, allowing it to measure the carnage close to the beam pipe.

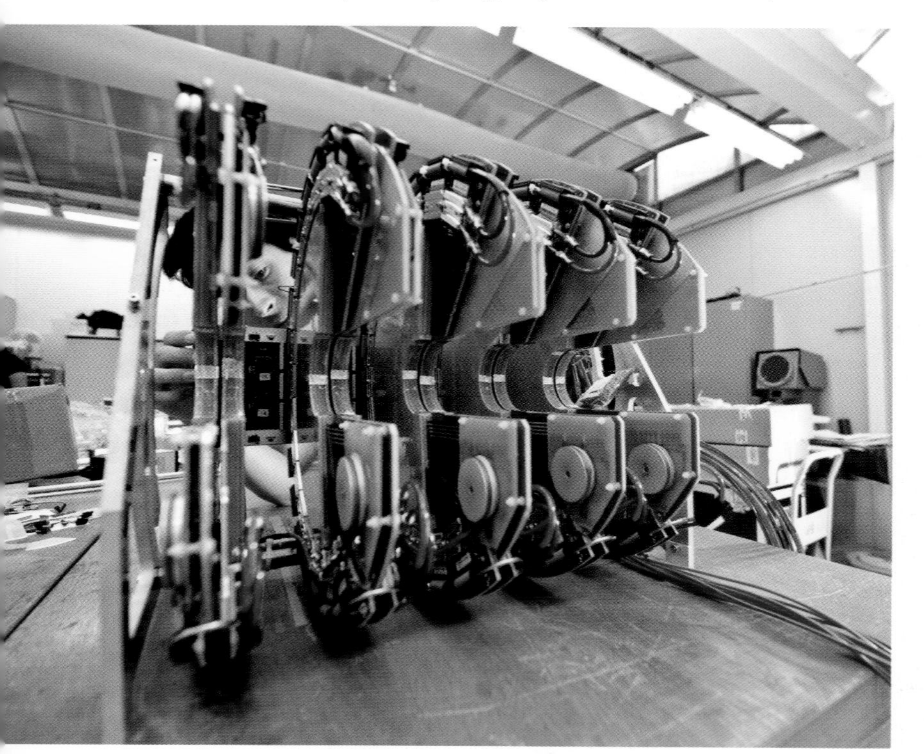

BELOW The TOTEM detectors. *(CERN/Mona Schweizer)*

You'll remember CMS' assistant in the Point 5 cavern, TOTEM? The smaller experiment would also help cover the collisions using its complementary software. 'Combining information from both allowed us to do a lot of physics studies that previously were impossible to do,' says Christof. 'It was possible to correlate proton remnants seen in TOTEM with objects such as jets and Upsilon particles observed in the central part of CMS.'

There was no denying that when lead ions met protons in the Large Hadron Collider there was some surprising behaviour to be had. In fact, particle physicists were arguing that they'd created a completely new type of matter, a colour-glass condensate that had originated out of two million lead-proton collisions. That's not all, though: the CMS team witnessed particle pairs flying away from each other, their directions in correlation. 'Somehow they fly at the same direction even though it wasn't clear how they communicate their direction with one another. That had surprised many people,' said physicist Gunther Roland at the Massachusetts Institute of Technology (MIT), United States.

It was true that Roland and other physicists had seen the condensate before – in the collisions between protons, where a liquid-like wave of gluons existed. But they weren't expecting it this time. In fact, they could only really explain the behaviour when protons were at work. It turns out that when a proton finds itself at higher energy levels, gluons cluster to its already existent quark trio. These gluons are able to exist as both particles and waves, their wave functions correlating with each other, explaining how they can share information and then decide to fly away from the impact.

Protons and lead ions continued to collide from January through to February of 2013, but it wasn't just CMS and TOTEM who were taking notes – LHCb and ALICE were also watching the action unfold; they too were keen to understand the properties of the strong interaction. In particular, LHCb was able to get a closer look at the making of strange, charm and beauty quarks in areas not accessible to the other experiments in the set-up of the Large Hadron Collider, despite its comparatively small size. Meanwhile, ALICE had found something exceedingly odd as it surveyed proton meeting lead. ATLAS had seen the same thing as well.

It was what CERN physicists called a 'double ridge', brought about by the particle wreckage in their results; it's exactly how it sounds – a plot that looks akin to a map of a mountain with lowlands and a ridge behind it. The teams behind the experiments realised that the structure was also made from heavy ions knocking together; they'd seen it when making the quark-gluon plasma. To some extent they were expecting the result, save for the ridge in the background. But it was odd that it was happening now.

What it meant exactly was that physicists had nothing to blame it on, other than it possibly being a blip in their data. Ridges in plots usually come about when a couple of particles fly off from each other at close to the speed of light in one direction, yet are positioned in another. Think of two tennis balls flying towards each other along one axis, then rebounding along another axis.

No matter how hard they tried, physicists

BELOW TOTEM being installed. *(CERN/Mona Schweizer)*

weren't successful in killing the feature off. What's more, the similar effect had been seen at the Relativistic Heavy Ion Collider, where physicists put it down to an effect from the creation of hot, dense matter. Could it be a colour-glass condensate predicted by quantum chromodynamics, which describes the strong interactions between quarks and gluons? Or could it be that the intensity between the protons and lead ions created some near-perfect fluid not too dissimilar to what detectors witnessed when lead ions crash into each other? Whatever it is, CERN were not prepared to settle for the same argument. It was a conundrum that was as exciting as it was frustrating.

What it could mean, however, is that there might be some profound implications from obliterating heavy ions at breakneck speeds. As it stands, though, the jury is still out.

LHCb meets new particles

The LHCb experiment was also finding itself at the centre of anomalies throughout the months that followed. The experiment had discovered that the B-meson seems to fall apart in a way that the Standard Model wasn't letting on about – the decay of particular interest being a kaon, a muon and an anti-muon during the crashing of protons. It was a deviation from a particle physics model, but the experiment's team couldn't deny their confidence. They pegged it as a 4.5-sigma, close to the 5-sigma required for a discovery.

There was a theory, though, that could be explained by the building blocks of particle physics. The decay could be down to the subtle effects of the W and Z bosons, the carriers of the weak force, at play. Any particle not described by the Standard Model could be contributing to the decay, which, in turn, is picked out by LHCb – in other words it would be brand new physics. But not everyone was convinced: 'I for one would not bet my house on this being a first sign of new physics, but it is certainly very interesting,' says Tim Gershon, who is on the LHCb team. He was being highly sceptical; others, not so much – they couldn't keep a lid on their excitement. What if it hinted at the possibility of a new particle, the Z boson? Physicists had theorised that it would be more massive than its namesake. But as ever, data needed to be grabbled before the find was a dead cert.

While data continued to be collected, analysed and stitched together to get a picture that made sense, the team behind the LHCb were kept more than occupied as it continued to churn out more and more findings. It had cornered two new heavy baryons by November 2014 – these were what are known as cascade particles, or the Xi baryon, a concoction of three quarks, usually of the up or down variety, coupled with a pair of massive strange, charm or bottom quarks.

Particle physics had already predicted their existence, but that didn't overshadow the finding. They'd found two particles for the price of one that are roughly six times heftier than the proton and were known as Xi_b* and Xi_b'. Being cascade particles, they don't live for very long – only a thousandth of a billionth of a second to be exact, before breaking up into pieces – components that the LHCb latched on to, using their momentum and mass to uncover the baryon pair.

Though they might be made up of exactly the same kind of quark, there are nevertheless some differences between the two. Their masses are slightly different, with Xi_b* being slightly heavier than its companion. Why this should be is all down to how the spins of two quarks are aligned; when they line up extra

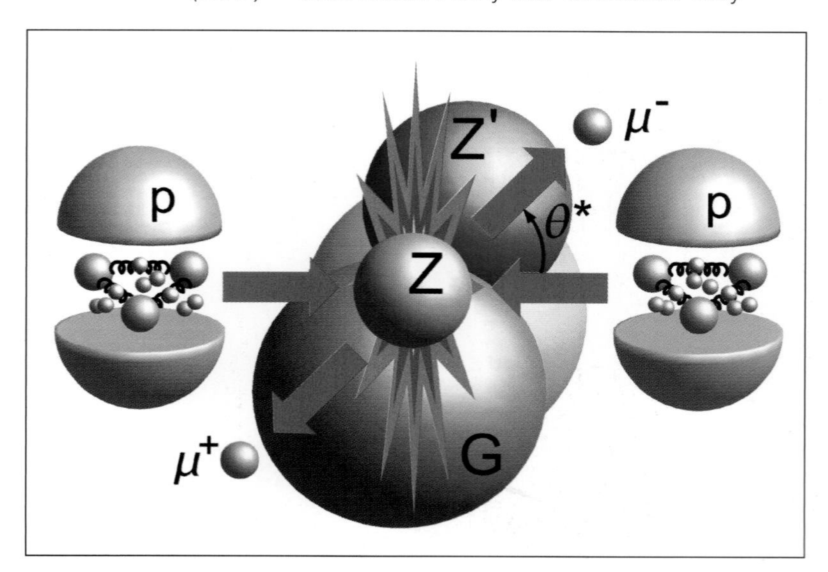

BELOW An artist's impression of the proton-proton collision that lead to the formation and subsequent decay into two muons of a newly discovered particle called a Z boson. *(CERN)*

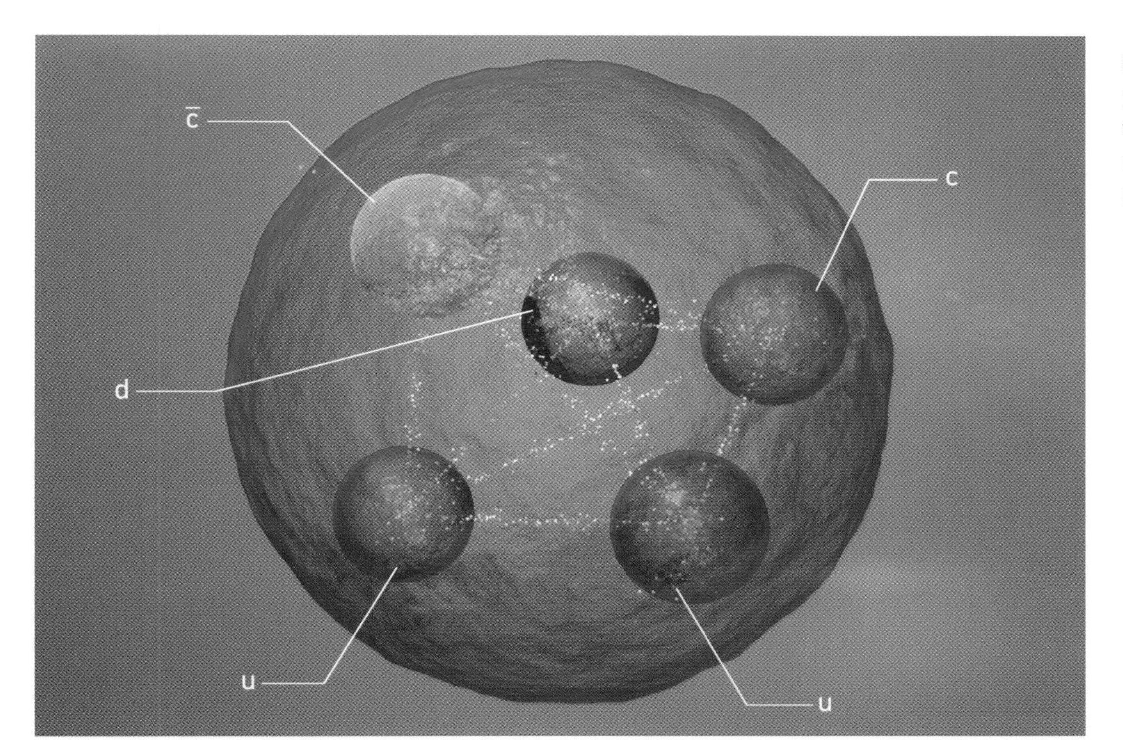

LEFT Another unusual particle discovered by the LHCb, a five-quark particle known as a pentaquark. *(CERN)*

energy is made, giving the baryon more mass – that's what's happened in the case of Xi_b*.

Such was its sensitivity that the LHCb experiment was turning up much more than physicists bargained for – not that they were complaining. A pentaquark had turned up; an exotic particle made up of five quarks. 'It's not just any new particle,' said Guy Wilkinson of the LHCb team. 'It represents a way to aggregate quarks, namely the fundamental constituents of ordinary protons and neutrons, in a pattern that has never been observed before in over 50 years of experimental searches. Studying its properties may allow us to understand better how ordinary matter, the protons and neutrons from which we're all made, is constituted.'

The announcement had been made during July 2015, a mere few months after the Large Hadron Collider had been powered up to fire proton beams around its tunnel at 6.5TeV apiece, followed by a collision with an energy of 13TeV – a new record for the particle smasher. Researchers had been looking at the decay of a baryon known as Lambda b, which in

BELOW Experiments in the LHCb taking place at a world record 13TeV. *(CERN)*

less than the blink of an eye had dissolved into a J/psi meson, a proton and a charged kaon. Checking their spectra built from LHCb data, they noticed a peak that gave the game away – a pentaquark was as clear as day. 'Benefitting from the large data set provided by the Large Hadron Collider, and the excellent precision of our detector, we had examined all possibilities for the signals and concluded that they could only be explained by pentaquark states,' announced Syracuse University College of Arts and Sciences' Tomasz Skwarnicki, who is a long-time member of the experiment's collaboration. 'The states must be formed of two up quarks, one down quark, one charm quark and an anti-charm quark.'

But there was still more the pentaquark could tell us. For one thing, how each of their quarks are bound together, teaching us a thing or two about Nature's interactive forces on a more subatomic level. 'The quarks could be tightly bound, or more loosely bound in a sort of meson-baryon molecule in which the baryon and meson feel a strong force just like one binding protons and neutrons to form nuclei.'

A weasel in the works

As the Large Hadron Collider continued to tackle the myriad mysteries of the universe, it had a visitor. Someone had taken a shine to one of the collider's high-voltage 66kV transformers: it was none other than a stone marten, which had wandered in from the surrounding countryside. The furry critter had climbed on top of the unit not too far from the LHCb experiment and gnawed its way through a power cable. The world's most powerful scientific instrument was brought to a complete standstill overnight.

When it came down to it, the animal was never going to be a match for the sheer electrical gravitas of the Large Hadron Collider. Picking up the charred remains of what they initially thought was a weasel, physicists couldn't believe the timing. They were preparing to lock down some new data on the Higgs boson and they'd also come across some tantalising hints of potentially new particles. 'We had electrical problems,' said head of press for CERN Arnaud Marsollier. 'It wasn't the best week for the Large Hadron Collider!' Nor would it be the last time it would happen either – just a few months later another marten literally collided with the machine, meeting a similar fate to its late cousin. 'These sorts of mishaps are not unheard of,' said Marsollier. 'We are in the countryside, and of course we have wild animals everywhere.'

Repairs took a few days, whilst getting the collider back up to its particle-smashing self again would take just over a week. When the machine was up and running, physicists would soon discover that the hints of particles the detectors had uncovered recently weren't to be taken lightly.

CERN finds four tetraquarks

The summer of 2016 saw the announcement of three kinds of tetraquark, which – as it sounds – is an exotic particle made up of four quarks. These new particles didn't represent any new forces, nor any interactions. They didn't solve any problems in the Standard Model. We'd just never seen them very often. It was also a long-standing theory that hadrons could only be made up of either a quark-antiquark pairing or three quarks. Just a couple of years previously we had seen confirmation of the existence of tetraquark Z(4430). Now it was going to have company.

It was the LHCb that sniffed the exotic particles out, known to be X(4140), X(4274), X(4500) and X(4700), the latter having higher masses. 'Even though all four particles contain the same quark composition, each of them has a unique internal structure, mass and set of quantum numbers,' says Skwarnicki. As you have discovered in Chapter 1, quantum numbers describe a particle's subatomic properties.

The tetraquarks that LHCb had found were discovered to be made up of two charm quarks, grouped with a couple of strange quarks. As subatomic particles go, they're the most massive of the quark family. 'The heavier the quark, the smaller the corresponding particle it creates,' says Skwarnicki. 'The names [of the new particles] are denoted by megaelectronvolts, referring to the amount of energy an electron gains after being accelerated by a volt of electricity.'

X(4140) isn't quite a new particle on the block, though – evidence for it had initially appeared back in 2009, at the Fermi National Accelerator Laboratory a stone's throw from Chicago. But it needed to be confirmed, thrusting the LHCb into the spotlight once more three years later, when Skwarnicki, clutching the results, announced that out of a whopping ten million proton smashes per second, the Large Hadron Collider's detector had seen the tetraquark – weighing in at four times the mass of the proton – almost 600 times. Increased sensitivity of LHCb's instruments would uncover X(4140) tetra-particle siblings.

Skwarnicki had seen clear evidence that their newfound particles were falling apart, disintegrating into a J/psi meson – which comprises a charm quark and its antiparticle – as well as a phi meson, which, as you'd expect, comprises a strange quark and its partner, a strange antiquark. 'We looked at every known particle and process to make sure that these four structures couldn't be explained by any pre-existing physics,' says Skwarnicki. 'It was like baking a six-dimensional cake with 98 ingredients and no recipe – just a picture of a cake.'

Indeed, interpreting datasets isn't all smooth sailing. In 2016 physicists were sure they'd found another particle – after all, the ATLAS and CMS experiments were in agreement. Crashing protons had brought suggestions of a brand new mysterious particle, an excess energy of 750GeV seemingly pointing the way to a brand new member of the particle zoo. In the end, however, it became 'the particle that wasn't', as more experimental tests at the Large Hadron Collider caused the bump in their results to flatline. It was a disappointment, but spokesperson for CERN Tiziano Camporesi couldn't deny the hype. 'It was disappointing, but the experimenters had always cautioned that the bump was most likely a fluke.' He added: 'We have always been very cool about it. This is the success of science, this is what science does.'

That's the attractive catch with particle physics; there's always more to figure out. The Large Hadron Collider and its team of experiments are at the forefront, there to fill in the gaps and make sense of the vastness of the mysterious cosmos that unfolds before us.

BELOW Another new particle discovered by the LHCb and featuring two heavy quarks (c), called the doubly heavy-quark baryon. *(CERN)*

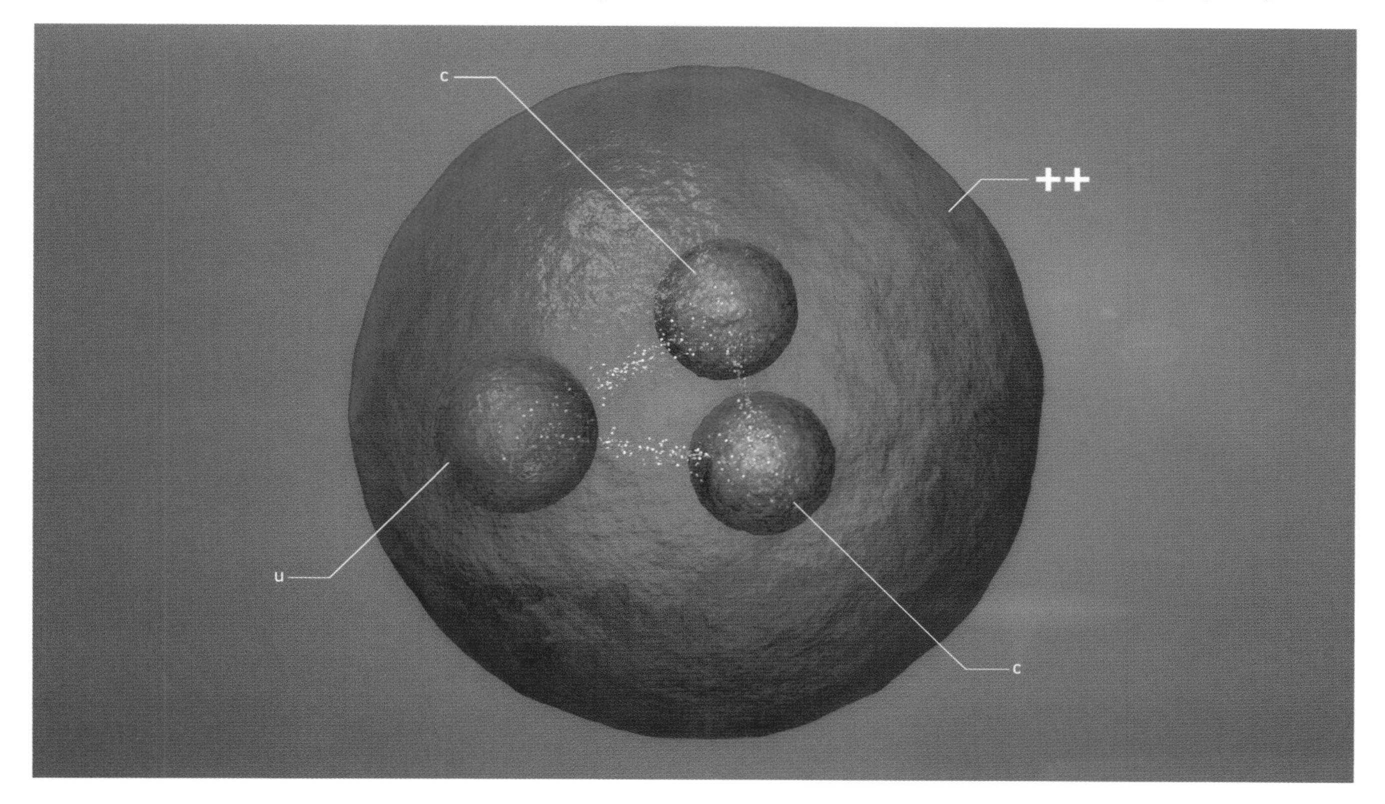

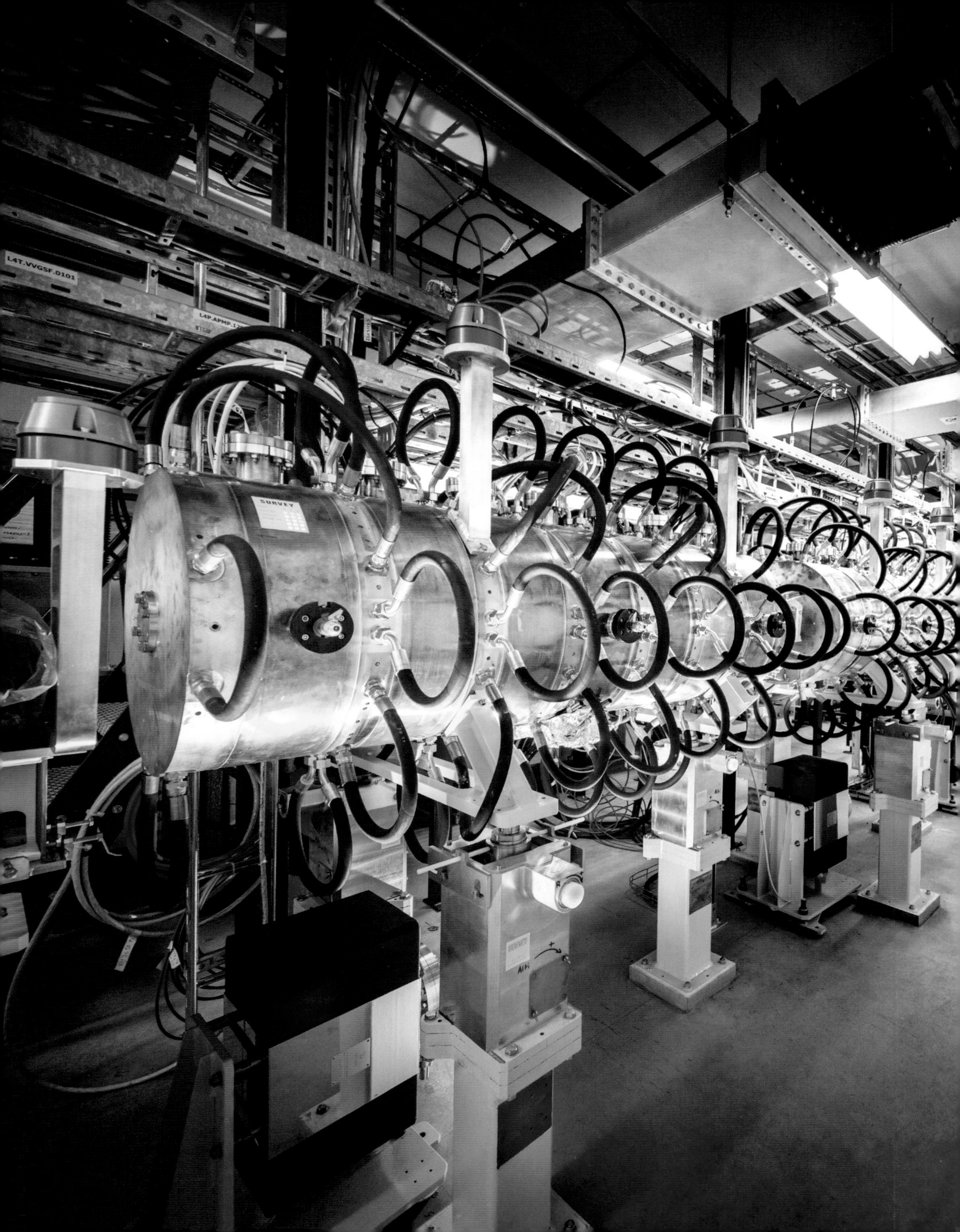
L4T.VVGSF.0101
SURVEY

Chapter Five

The future of the Large Hadron Collider and particle physics

The Large Hadron Collider may have achieved its main objective in following the trail of breadcrumbs to the long-sought Higgs boson, but CERN weren't looking at shutting up shop and going home just yet. There were still more mysteries of the universe to figure out.

OPPOSITE Injector LINAC 4 is the first key element in the upgrade programme of the Large Hadron Collider. *(CERN)*

From the moment it first sprang to life, to its exotic collisions of particles and its ramped-up luminosities and energy, the Large Hadron Collider has finally hit its maximum peak, smashing particles at the intensity for which it was built. For every powering down for the winter, the particle smasher has been moved to the next best thing, rolling on to the next version – from top subdetectors to the latest interfaces for its computer system. 'The control systems [in the control rooms] are Windows, but the majority of experimental systems are Red Hat Linux. We've migrated from scientific Linux to CentOS,' says Arturo Sánchez Pineda, a physicist behind the world's most powerful experiment. 'Security patches are also important and we have a custom C++ [computer programming language] framework to analyse the data. You could save it, maybe as an Excel table or something, but it would be incredibly big.'

There's no denying that the collider spits out a lot of data. But the scientists behind it believed that the particle smasher could always do with being bigger. And it could always do with being better. By at least ten times its design value, in fact.

And they have a plan. Enter the High-Luminosity Large Hadron Collider, which is exactly as its sounds: a souped-up version of the particle smasher, designed to probe even further into the Standard Model and beyond. Of course, the accelerator isn't being redesigned, it's simply getting a few add-ons and upgrades that'll help it to get into those tough-to-reach places, still locked in the realms of particle physics.

As you'll gather from its exotic title, the name of the game with this particle smasher is to jack up the luminosity, something which, as you'll have already noted from reading this book, is an important factor in the world of accelerators. It tells us something about the number of collisions that happen in a given moment, and the higher the luminosity, the more data its detectors can snatch. The more they see, the more likely they are to find some kind of rare process in the subatomic world.

Scientists first uttered the upgraded machine's name during a special meeting in Brussels. Their gathering was hosted by the European Commission, and the High-Luminosity Large Hadron Collider had top billing on the agenda. 'The design of the [accelerator] was to reach 14TeV,' explains Pineda. 'The machine has been working very well, so everyone has the idea that we can push past that.'

Just for reference, the High-Luminosity Large Hadron Collider will be able to make 15 million Higgs bosons a year when it's up and running by 2025, compared to the 1.2 million witnessed by the experiments on the particle accelerator's current circuit throughout 2011 to 2012. But to get from the world's most powerful collider to super collider there need to be some adjustments – something that began in earnest not too long after the attendees of the European Strategy for Particle Physics parted ways.

BELOW A short-model magnet for the High-Luminosity Large Hadron Collider's quadrupole. *(CERN/Robert Hradil/Monika Majer)*

Upgrades in the background

From previous chapters you'll remember that the particle accelerator undergoes annual shutdowns to give engineers and scientists the chance to regroup, talk about results and carry out all-important upgrades and fixes to magnets, its range of experiments and more. Behind the scenes, though, there was more than met the eye; CERN had also set the wheels in motion for the High-Luminosity Large Hadron Collider, and were busy with the next major development in superconducting technology. The particle smasher was getting

LEFT These brand new quadrupole magnets will focus particle beams into collisions for the High-Luminosity Large Hadron Collider. *(CERN)*

high-field superconducting magnets and even more compact, even-more-precise radio frequency cavities, as well as 300 million superconducting links that would allow the delivery of higher power. Superconductivity is what brings a particle accelerator to life, and these new additions would transform the machine, manipulating and bending particle beams to bring them to their high-energy smash even more effectively.

First there's the magnets, based on niobium-tin superconductor. They're capable of bringing about stronger magnetic fields than standard niobium-titanium mix structures. They've also recently been tested in the United States at the Fermi National Accelerator Laboratory, commonly known as Fermilab, which culminated in the unveiling of a 10.4-tesla, 2m prototype magnet. 'The idea originated from a proposal made by Lucio Rossi, who was the head of CERN's Magnets, Superconductors and Cryostats group, back in 2010,' says the head of Fermilab's Technical Division, Giorgio Apollinari. 'During a discussion he suggested replacing a few of the Large Hadron Collider's 8-tesla dipole magnets with shorter 11-tesla magnets.' The particle accelerator's 27km ring was already full to bursting point, so, in order to slot in some extra collimators to narrow the

BELOW Superconducting links have been developed to carry currents of up to 20,000A ready for the particle smasher's upgrade. *(CERN)*

ABOVE A model of a Racetrack Model Coil (RMC), which is being developed in order to understand and develop new technologies vital for future accelerators. *(CERN)*

beams into the set-up, reducing the size of the magnets was the answer.

So that's what the magnets needed to be – small but mighty. This led to further phases of working and reworking their design until both CERN and Fermilab were happy with the product of their collaboration: 11m magnets, 3m shorter than the ones you'll find already in place at the Large Hadron Collider. 'Niobium-titanium was the material used in the manufacture of the superconducting cables of the [accelerator's] magnets in the 1990s,' says Apollinari. 'We decided to use niobium-tin instead.'

BELOW Just one of the first crab cavities being constructed in a clean room at CERN, which will assist with increasing the luminosity of the Large Hadron Collider. *(CERN/Ulysse Fichet)*

Before they could even think about slotting them into the Large Hadron Collider's set-up, further testing was needed to truly understand the new magnetic technologies. Teams at CERN had a device, a Racetrack Model Coil (RMC), that could help. They were interested in seeing just how far they could push the niobium-tin construction of the newly crafted magnets – despite their brittle nature – if it ever came to it.

A world record was broken close to the end of 2015, when engineers gave their test magnet a whirl. They had whacked the field up to 16 tesla, almost twice what coursed through the Large Hadron Collider. 'It was an excellent result,' admits Luca Bottura, Head of CERN's Magnet Group, 'although we couldn't forget that it is a relatively small magnet, a technology demonstrator with no bore through the centre for the beam. There's still a way to go before 16-tesla magnets can be used in an accelerator. Still, this is a very important step towards them.'

Next, there's what are known as – due to their appearance – 'crab cavities' for beam tilting, which cause the particles to fall sideways in a motion that's akin to the scurrying of such crustaceans. Once the particle bunches are made, they need to be slanted in such a way that they crash into each other at specific points. There will be 16 of these robotic crabs, lying in wait close to ATLAS and CMS, ready to use their superconducting properties to ensure the proton bunches meet for a high-luminosity run of the Large Hadron Collider. CERN is currently the proud owner of two of these cavities, safely locked away in a cryostat that keeps them chilled at a temperature of 2°K, which equates to about -271°C (-456°F). That's how they're able to operate at their very best. They'll be packed into the Super Proton Synchrotron when the time comes to see how they fare with their very first packets of protons.

These high-speed protons are spat out by the refurbished injector chain, four accelerators that get the beams up to speed before they enter the Large Hadron Collider's main domain. You'll remember the LINAC 2 injector, which is where a particle's journey begins before it's

RIGHT A vertical cryostat is tested before being installed as part of the High-Luminosity Large Hadron Collider, which aims to crank up the performance of the collider. *(CERN/Maximillien Brice)*

accelerated to speeds close to that of light. It's been working for CERN since 1978, but the summer of 2017 saw murmurs of leading the Large Hadron Collider's chain into retirement. CERN want to bring in the big guns with LINAC 4. 'LINAC 4 is a modern injector and the first key element of our ambitious upgrade programme, leading us to the High-Luminosity Large Hadron Collider,' says CERN Director General Fabiola Gianotti. 'The phase will considerably increase the potential of the Large Hadron Collider experiments for discovering new physics and measuring the properties of the Higgs particle in more detail.'

It's taken almost ten years to build LINAC 4. It's 90m in length and has the power to shape the density and intensity of the particles it sculpts into beams. Its weapon of choice is the negative hydrogen ion, which it hands over to the Proton Synchrotron Booster to rev up the energy beyond 160MeV; that's more than three times the energy LINAC 2 is able to achieve. The engineers at CERN have been tactful in their choice of hydrogen ions. They figured that combining them with skyrocketed energy is the perfect recipe for a beam intensity which is double that delivered by the Large Hadron Collider. That's where LINAC 4 is contributing to a jacked-up increase in luminosity.

Of course, you can't have a brand new injector without a new racetrack on which the beams can get up to speed; consequently the particle physics laboratory has made the decision to give the Proton Synchrotron Booster, the Proton Synchrotron and the Super Proton Synchrotron a bit of a spit and a polish too. Just recently the latter cog in the works had a beam of partly ionised xenon particles flawlessly injected into it, just before the otherwise inert gas particles made their way inside. You'll find xenon – a colourless, dense, odourless gas – in the Earth's atmosphere, as well as in lamps. But this time each and every atom in the particle accelerator has 39 of its 54 electrons shaved off.

BELOW New sensors for the High-Luminosity Large Hadron Collider will replace the Compact Muon Solenoid's end-cap. *(CERN/Maximilien Brice)*

That's what makes them quite fragile, so it's a massive feat to get this ionised gas circulating before they have a chance to disintegrate. Losing just one electron from its configuration causes an atom to become erratic, changing its direction and going wherever the mood takes it. But CERN do have a helping hand when it comes to the more delicate members of the Large Hadron Collider's particle beam family: a vacuum. In fact, the gas serves kind of like a cushion, but CERN has readily admitted they haven't quite got the technique of handling xenon beams with kid gloves down pat just yet. 'The Super Proton Synchrotron vacuum is not quite as high as that of the Large Hadron Collider,' explains Reyes Alemany, who's responsible for testing out the Super Proton Synchrotron. 'Keeping the beam going for one cycle is already a promising result.'

The reason why CERN are juggling xenon atoms is to get to grips with the basics in building a gamma ray factory. They want a high-intensity source of photons with ultra-high energies of up to 400MeV, not too dissimilar to the intensity of synchrotrons. 'The source would pave the way for studies never done before in fundamental physics, like in the fields of dark matter research,' explains physicist Witold Krasny. 'It also opens the door for industrial and medical applications and could possibly serve as a test bench when making a future neutrino factory or muon collider.'

Every test, every novel idea, that CERN has in getting ahead in the particle physics game paves the way to the Large Hadron Collider's upgrade. In fact, it's quite easy to get lost in beam intensities and the latest advances in magnet technologies, but there's more digging to be done in the tunnels, and up to 80 brand new collimators will need to be fitted to ensure the Franco-Swiss particle collider can handle the high-luminosities. After all, there will be 140 collisions every time particle bunches meet right in front of the watchful, unblinking detector eyes of ATLAS and CMS. They're currently taking it easy with 30 subatomic smashes. There's another advantage to increasing the luminosity: when protons collide they won't vanish into thin air, plus there'll be roughly the same amount every time the particles go head to head. CERN is also looking to squeeze the beams into smaller threads, with the help of an Achromatic Telescope Squeezing (ATS) scheme.

It's intended that a couple of new tunnels will be carved out quite close to the cavernous homes of ATLAS and CMS. They'll be about 300m in length and will be the place where CERN slots in equipment that's sensitive to radiation; physicists are, of course, thinking of the power converters that'll be capable of switching alternating current sucked from the collider's electrical network into a direct, high-intensity flow of charge to juice up the magnets. The converters will be linked up to the High-Luminosity Large Hadron Collider via superconducting transmission lines, their magnesium diboride tubing effortlessly siphoning currents of 100,000A and operating at freezing temperatures of -235°C (-391°F). They'll share the same cavern as a cryogenic kit that'll keep everything running at just the right temperatures, just under two brand new shafts – about 100m deep – that are to be excavated so that engineers have easy access to the service tunnels beneath their feet.

There's no time like the present when it comes to getting everything in place to see the High-Luminosity Large Hadron Collider up and running. Rewind back to when the Large Electron-Positron collider was in operation, with the Large Hadron Collider waiting in the wings, ready to be installed. CERN are in the same position now, only they don't want to disturb the Large Hadron Collider too much in the upgrade to make it a better, high-luminosity version of itself. So to ensure that they don't upset it, teams of engineers have been measuring vibrations where the ATLAS experiment is installed, at Point 1 on the particle accelerator's ring. While the Large Electron-Positron collider could withstand a touch of shaking of the earth around it, its successor is much more sensitive to vibrations. It's a tricky task, but someone's got to do it.

'While the main civil engineering work will, of course, take place during the long shutdown scheduled for July 2018, we're looking to identify which parts of it could be carried out during the Large Hadron Collider's operation,' explains Paola Fessia, a physicist at CERN who's dealing with the installation of the souped-up particle accelerator. He realises that they'll need to get within 40m of the beam, so he and his team have installed several sensors to get a clearer picture of the vibration before a core-drilling machine arrives. 'The first vibrations we studied were generated by a core-drilling machine used to examine the site's geological make-up. This information will be essential for designing and constructing the new underground caverns and technical galleries needed for the High-Luminosity Large Hadron Collider, as construction companies need to know exactly what they'll find when they dig into hard rock, water and sand. This is the main purpose of the drilling. It has also been

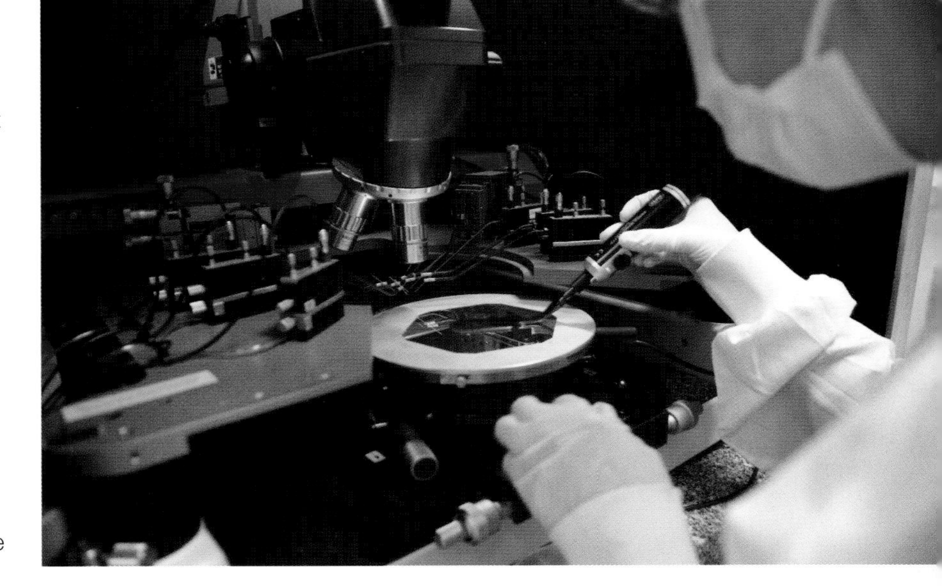

ABOVE Sensors are carefully placed into a probe station. They're being tested for the High-Luminosity Large Hadron Collider by applying a high voltage through them using a needle. *(CERN/Ulysse Fichet)*

BELOW A seismic truck arrives at Point 1 to generate vibrations to test the geological make-up and work out if upgrades can continue on the Large Hadron Collider during its operations. *(CERN/Sophia Bennett)*

used to study the effect of pulsed vibrations.' A 24-tonne machine arrived next. This was the seismic truck, which forced its weight into the ground to stir up tremors at about 100 times per second. 'We created waves with a wide range of frequencies,' says Michael Guinchard, chief of the mechanical measurement laboratory.

The future of the world's greatest particle collider

'It's only prudent to try to sketch a vision decades into the future,' says theoretical physicist Michael Peskin, currently based at the SLAC National Accelerator Laboratory in California.

BELOW Aerial view of the SLAC National Accelerator Laboratory in California. *(SLAC)*

If your particle smasher isn't quite as sensitive as you'd like, or it's no longer hitting the sweet spots of discovery, then what do you do? As we've learned up to this point, CERN are always looking to spearhead the latest technological advances, to ensure that the Large Hadron Collider is always at peak performance. After all, it's not called the world's most powerful particle accelerator for nothing.

Sometimes, though, it makes more sense to build an even bigger particle collider. That's where the Future Circular Collider comes in, which has also been given the name of the Very Large Hadron Collider and is tipped to generate such a leap in energy that it could find particles heavier than the Higgs, giving us an even deeper insight into the laws that keep the cosmos tied together. 'What are the details of the Higgs? And what is the dark matter particle, or particles?' muses particle physicist Geoff Taylor, who is based at the University of Melbourne in Australia.

It's true that, with its 27km particle circuit, the currently operational hadron smasher is the lord of the rings, but those looking to construct its next iteration want to go bigger, up to 100km (62 miles) around, thereby allowing particles to be smashed at an eye-watering 100TeV – that's the equivalent of ten million lightning strikes. From its roughly mapped-out position that would see it snake partially under Lake Geneva, it would seek out the building blocks of Nature like never before. It's hoped that this new addition will be in place by the end of the 2030s, not too long after the High-Luminosity Large Hadron Collider brings its beams into a collision for the very last time.

And the Large Hadron Collider wouldn't be given the boot either. It would come along for the ride in the future of particle physics, revving up particle beams just like its synchrotron predecessors and feeding them to the Future Circular Collider, the High-Luminosity Large Hadron Collider. CERN's Frank Zimmermann, one of the brains behind the design of the Future Circular Collider, is cautious of the inevitable challenges that will confront the assembly of such a monstrous machine.

Firstly, there's the major hurdle of deciding if a 16-tesla magnet can actually be made in such a way that's not going to break the bank. 'It's a major step compared to the 8.3-tesla magnets that are currently being used at the Large Hadron Collider, as we will have to double the magnetic field,' explains Zimmerman. 'We also have to consider how to minimise the cost for operating these superconducting magnets to make the project feasible.'

Secondly, there are the enormous wads of energy that'll be packed into the particle accelerator's beams. They need to be kept under control. 'We would like to avoid uncontrolled beam loss,' says Zimmerman. 'If the entire beam were lost somewhere in the arcs this would be catastrophic and could cause expensive damage by destroying numerous magnets and other parts of the infrastructure.' Of course, work on the Large Hadron Collider has proved useful in this sense; in fact, CERN have already had 20 years of beam control. 'But there are still steps forward to be made,' says Zimmerman. And that's not all. A 100TeV proton-smasher will generate an enormous amount of synchrotron radiation. It'll also create a lot of heat that needs to be absorbed from the inside of the cold magnets.

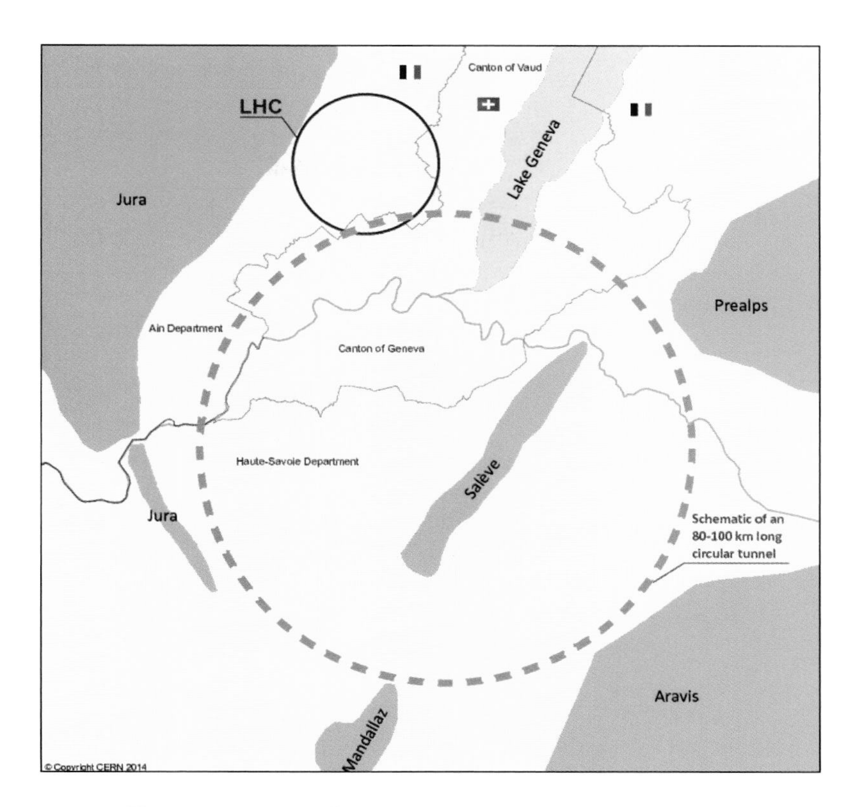

ABOVE The Future Circular Collider will run for 100km, using the Large Hadron Collider's tunnel and magnets to ramp up the particles to speeds close to that of light. *(Emilie Ter Laak)*

That's just some of the issues they'll encounter. There's also the matter of what type of collider will fill and exceed the shoes of the Large Hadron Collider. Accelerator physicists have three ideas in mind: the Future Circular Collider could take after its predecessor, pumping protons and hefty ions around its network of magnets; it could take after the Large Electron-Positron collider, sending electrons and positrons into an almighty smash; or it could take inspiration from both colliders and use a flavouring of proton and electron particle beams.

In terms of areas of the Standard Model physicists can probe with a future collider, they're spoilt for choice. A more supercharged version of the Large Hadron Collider will dig deep into revealing carriers of new forces in the cosmos, allowing dark matter to be pulled apart just enough to examine its constituents and make the supersymmetric partners of gluons and quarks, whilst simultaneously finding out if the latter have some kind of substructure we don't know about. It'll also make billions of Higgs bosons and the top quark, sniffing out more unusual decays at enormously high collisional energies. Meanwhile, an electron-positron accelerator with a high luminosity can offer something similar, whilst delving deep into the properties of the W and Z bosons with unrivalled accuracy. Go for an electron-proton collider and you're looking at a Higgs boson factory, an experiment that would serve as a microscope that would shake loose all kinds of particles and other such matter. For accelerator physicists, there are endless possibilities.

When it comes to building, it's a bit more plain sailing if they go for the lepton-collider set-up. We have a lot of the technologies in place for such an accelerator already, including superconducting radio-frequency systems, and we'll need the magnetic fields coursing through the new accelerator to be pretty low, at, say, 0.4 teslas. Of course, creating magnets of such a requirement could wind up being costly. Zimmerman doesn't seem to be fazed, though: 'I think in the coming years we will be able to master this technology and show significant results,' he says.

Compared to the Large Electron-Positron collider and its successor, the future collider is aiming high; it's looking to dazzle with a luminosity somewhere between ten and 100 times greater. 'Compared to the Large Electron-Positron, the Future Circular Collider will have a larger number of bunches and two separate beam pipes are foreseen,' explains Zimmerman. 'We also need to demonstrate that the footprint of the lepton collider can be equal to one of the hadron collider, so that both can fit in the same tunnel.' He is referring to the challenge that engineers will come across in controlling the strong synchrotron radiation; bending the beams too much and around the interaction hotspots will create irritating background noise for the experiments on the collider's ring.

If Zimmerman and his team are looking to take the hadron-lepton collider route, then electrons will need to be spat out by an energy recovery linear accelerator. Better still, physicists could employ a ring on the Future Circular Collider that carries just protons and another that streams electrons. 'This would require sufficient space in the transverse tunnel cross-section to host both the lepton and hadron rings,' says Zimmerman. 'We've yet to consider it.'

Regardless of which design is on the cards in the future, there's a wealth of information to be gleaned from building it, even on a path beyond its usual jurisdiction. 'On the one hand, it's asking those very fundamental questions, but on the other hand, it's not forgetting that there is almost always a direct link to applications that benefit society immediately,' says head of physics Carsten Welsch, who's at the University of Liverpool. After all, dabbling in databases and software of computers at CERN in the lead-up to 1989 brought about the birth of the World Wide Web through the research of computer scientist Tim Berners-Lee.

The Future Circular Collider, as with every particle accelerator before it, will offer some headway into benefiting our lives on Earth through development of radiation-resistant materials useful in future nuclear reactors and power networks. Its high-field magnets can also assist with technologies like Magnetic Resonance Imaging (MRI) scanners, improving on their resolutions by trying out an increase in magnetic field strengths. 'Accelerators all over the world are performing a large variety of important functions,' adds Zimmerman, 'including [the creation] of tumour-destroying beams to fight cancer, killing bacteria to prevent food-borne illnesses; production of ever more powerful semiconductor devices and computer chips; examining cargo for homeland security and helping scientists improve fuel injection to make more efficient vehicles.'

One of the many experts wanting to see the Future Circular Collider succeed in a post-Large-Hadron-Collider-led era is Nima Arkani-Hamed, a leading particle physicist and director of the Centre for Future High Energy Physics in Beijing, China. 'One thing about being a professional scientist is not to ask the big question,' he said. 'It's to ask the next question.'

And, with the physics version of the discovery of DNA under its belt when it snared the Higgs boson amongst its other tentative steps in unlocking the secrets of the cosmos, the Large Hadron Collider will be ready and waiting with the answer, underneath the border of two European countries but with the backing of the entire world.

BELOW The workstation used by Tim Berners-Lee when he first began developing the World Wide Web at CERN. *(CERN)*

Acronyms

A – Amps.
AA – Antiproton Accumulator.
ACORDE – ALICE Cosmic Rays Detector.
AdA – Anello Di Accumulazione.
ADONE – 'Big AdA'.
ALEPH – Apparatus for LEP Physics.
ATLAS – A Toroidal LHC ApparatuS.
ATS – Achromatic Telescope Squeezing.
CEDAR – Combined e-Science Data Analysis Resource.
CERN – European Organisation for Nuclear Research.
CHF – Swiss francs.
cm – Centimetres.
CMS – Compact Muon Solenoid.
COMPASS – Common Muon and Proton Apparatus for Structure and Spectroscopy.
CP – Charge Conjugation Parity.
DATE – ALICE Data Acquisition and Test Environment.
DCAL – Dijet Calorimeter.
DELPHI – Detector with Lepton, Photon and Hadron Identification.
DORIS – Doppel-Ring-Speicher.
EMCal – Electromagnetic Calorimeter.
EPA – Electron Positron Accumulator.
ESA – European Space Agency.
Fermilab – Fermi National Accelerator Laboratory.
FMD – Forward Multiplicity Detector.
GEM – Gas Electron Multiplier.
GeV – Gigaelectronvolts.
Gy – Grays.
HCAL – Hadron Calorimeter.
HeRSCHeL – High Rapidity Shower Counters for LHCb.
HF – Hadronic Forward detector.
HI – Heavy Ions.
HIE – High Intensity and Energy.
HLT – High-Level Trigger.
HMI – Human-Machine Interface.
HMPID – High Momentum Particle Identification Detector.
HPD – Hybrid Photon Detector.
ISOLDE – On-Line Isotope Mass Separator.
IT – Inner Tracker.
ITS – Inner Tracking System.
kA – Kiloamps.
km – Kilometres.
kV – Kilovolts.
LabVIEW – Laboratory Virtual Instrument Engineering Workbench.
LEP – Large Electron-Positron collider.
LHC – Large Hadron Collider.
LHCb – Large Hadron Collider beauty.
LHCf – Large Hadron Collider forward.
LIL – Large Electron-Positron Injector LINAC.
LINAC – Linear accelerator.
MACHOs – Massive Astrophysical Compact Halo Objects.
m – Metres.
µ – Microns (see Micrometres).
µm – Micrometres (see Microns).
MeV – Megaelectronvolts.
MHz – Megahertz.
MoEDAL – Monopole and Exotics Detector at the Large Hadron Collider.
MW – Megawatts.
NA61/SHINE and NA62 – Heavy ion and neutrino experiments.
NASA – National Aeronautics and Space Administration.
OERN – Organisation Européenne pour la Recherche Nucléaire.
OPAL – Omni-Purpose Apparatus for LEP.
PEP – Positron-Electron Project.
PETRA – Positron-Electron Tandem Ring Accelerator.
PHOS – Photon Spectrometer.
PMD – Photo Multiplicity Detector.
PS – Proton Synchrotron; also Pre-Shower Detector.
RF – Radio frequency.
RHIC – Relativistic Heavy Ion Collider.
RICH – Ring Imaging Cherenkov detector.
RMC – Racetrack Model Coil.
SCT – Semi-Conductor Tracker.
SLAC – Stanford Linear Accelerator Center.
SPD – Silicon Pixel Detector.
SPS – Super Proton Synchrotron.
SSC – Superconducting Super Collider.
T0 – Time-zero detector.
TeV – Teraelectronvolts.
TIM – Train Inspection Monorail unit.
TOTEM – TOTal Elastic and diffractive cross-section Measurement.
TPC – Time Projection Chamber.
TRD – Transition Radiation Detector.
TRT – Transition Radiation Tracker.
TT – Trigger Tracker.
UA – Underground Area.
V0 – Vertex-zero detector.
V0-A and V0-C – Arrays of scintillator counters.
VELO – Vertex Locator.
WIMPs – Weakly Interacting Massive Particles.
WMAP – Wilkinson Microwave Anisotropy Probe.
ZDC – Zero Degree Calorimeter.

Glossary

ALICE – A Large Ion Collider Experiment, one of seven detectors inside the LHC.

Antimatter – Matter made from 'anti-particles', that is, particles with opposite electric charge and quantum numbers to their 'matter' counterparts. For example, an electron has a charge of –1, whereas its antiparticle, the positron, has a charge of +1.

ATLAS – A Toroidal LHC ApparatuS, which is the largest of the seven experiments inside the LHC.

Bosons – a boson is one of the two fundamental types of particle. They have integer or zero spins. Bosons can occupy the same energy levels and the same place in space as each other. Force carriers, such as the Higgs particle, are bosons.

Calorimeter – a device to measure the energy of particles.

CERN – The European Organisation for Nuclear Research. The acronym comes from the French name, Conseil Européen pour la Recherche Nucléaire.

Cherenkov radiation – the electromagnetic equivalent of a sonic boom, a flash of blue light emitted when a relativistic particle passes through a dense medium where the speed of light is slower than the velocity of the particle passing through.

CMS – Compact Muon Solenoid, which is one of the four big experiments at the LHC and is designed for general purpose physics.

CP symmetry – in physics, a symmetry refers to a property that remains unchanged following some physical transformation. CP symmetry involves charge conjugation, which permits a particle to become its antiparticle, and parity, which states that the mirror image of a particle system is unchanged.

Dark matter – a mysterious form of matter that is invisible and interacts only by gravity. It accounts for 85 per cent of all the matter in the cosmos.

Dipole magnet – a magnet with two opposite poles, north and south.

Electromagnetic force – one of the fundamental forces of nature, which governs interactions between charged particles.

Electron – a fundamental subatomic particle, with an electric charge of –1 and a tiny mass 1/1836 of a proton. Electrons orbit atomic nuclei to form atoms.

Electronvolt – a unit of energy equal to 1.6 x 10^{-19} joules. It is the energy lost or gained when an electron passes through an electric potential of 1 volt. Because Einstein taught us that mass and energy are the same thing, the masses of particles are often referred to in electronvolts (eV).

Fermion – these are the second fundamental type of particle with half-integer spins, such as –1/2, 1/2, 3/2. These include quarks, which make up protons.

Hadron – particles made from assemblages of quarks are called hadrons. Protons, neutrons and atomic nuclei are all hadrons.

Higgs boson – the Higgs boson is the force-carrying gauge particle of the Higgs field, which gives fundamental particles their mass.

High Luminosity LHC – an upgraded LHC, set to begin work in 2025.

Ion – an atom or molecule that has lost (or gained) one electron, giving it a net electrical charge, either +1 or –1.

Lepton – a form of fundamental particle that includes electrons, muons, tau particles and neutrinos.

LHCb – The LHC-beauty experiment, which looks for violations of CP symmetry and studies interactions of hadrons made with a 'beauty' quark.

LHCf – The LHC-forward experiment, which studies particles in the forward region of collisions, almost in line with the proton beams.

Magnet quench – the shutting down of a magnet when the magnet coil is no longer superconducting, caused by either the magnetic field itself, or the rate of change of the field, becoming too great.

Meson – a hadron composed of just two quarks, specifically a quark and an anti-quark.

MoEDAL – Monopole and Exotics Detector At the LHC, which is designed to search for magnetic monopoles, which have a single pole, and exotic particles such as dark matter.

Muons – these are leptons with the same charge and spin as an electron, but a mass over 200 times greater than an electron.

Partons – the point-like quarks that make up particles such as protons, and the gluons that bind the quarks together.

Proton – a hadron made of three quarks, two up and one down. Combinations of protons and neutrons (one up and two down quarks) form the basis of atomic nuclei.

Quadrupole magnet – a quartet of dipole magnets laid out in such a fashion that opposite poles of the quadrupole are the same.

Quarks – fundamental particles that are the basis for protons, neutrons, mesons and therefore almost all of the matter that we can see in the universe.

Quark–gluon plasma – a state of matter at high temperature and density, where quarks and gluons are free and not bound inside hadrons or mesons. A quark–gluon plasma is the most dense form of matter known outside of the Big Bang.

RHIC – the Relativistic Heavy Ion Collider at Brookhaven National Laboratory in New York. It is the second most powerful particle accelerator in the world.

Standard model – our picture of the fundamental particles and the forces that hold them together and allow them to interact. The Higgs boson was a crucial missing piece of the Standard Model.

Statistical level – the probability that an experimental result is correct can be given by the standard deviation away from the mean value of the dataset. 'Five-sigma' (five standard deviations) means that the likelihood that the data suggesting a specific result, such as the discovery of the Higgs boson, has a 1 in 3.5 million chance of being a fluke coincidence.

String theory – a popular theory that tries to unify quantum physics with general relativity by proposing that all matter, including fundamental particles such as quarks and electrons, is made up of tiny vibrating particles.

Strong force – the fundamental force that acts within the atomic nucleus, preventing similarly charged protons from repelling one another, and binding the quarks together inside those protons.

Superconductor – a material with zero electric resistance. A superconducting magnet is a magnet made from coils of superconducting wire. All known superconductors require extremely cold temperatures and so the magnets are chilled by liquid helium.

Supersymmetry – a theory that suggests that each boson and fermion particle in the Standard Model is linked to a fermion and boson 'super-partner', respectively. However, none of these super-partners have ever been discovered.

Tevatron collider – a particle accelerator at the Fermi National Accelerator Laboratory, which was shut down in 2011.

TIM – the Transport Inspection Monorail that searches the LHC tunnel for faults.

TOTEM – TOTal Elastic and diffractive cross-section Measurement, one of the smaller of the seven experiments in the LHC.

Index

LHC (and LEP) construction, components and operation 27-95

Particles and other matter

People